Bioreduction of selenite and tellurite by *Phanerochaete chrysosporium*

Thesis committee

Promotor
Prof. Dr. Ir. Piet N.L Lens
Professor of Biotechnology
UNESCO-IHE
Delft, The Netherlands

Co-Promotors
Dr. Hab. Eric D. van Hullebusch
Hab. Associate Professor in Biogeochemistry
University of Paris-Est
Marne-la-Vallée, France

Dr. Hab. Giovanni Esposito
Hab. Associate Professor in Environmental Engineering
University of Cassino and Southern Lazio
Cassino, Italy

Mentor
Dr. Eldon R. Rene
Senior Lecturer in Resource Recovery Technology
UNESCO-IHE
Delft, The Netherlands

Other Members
Prof. Dr. François Farges
Professor of Mineralogy
Museum Nationale d'Histoire Naturelle
Paris, France

Prof. Dr. Erkan Sahinkaya
Professor of Bioengineering
Medeniyet Üniversitesi
Goztepe, Istanbul, Turkey

Prof. Dr. Jonathan Lloyd
Professor of Geomicrobiology
University of Manchester
Manchester, United Kingdom

This research was conducted under the auspices of the Erasmus Mundus Joint Doctorate Environmental Technologies for Contaminated Solids, Soils, and Sediments (ETeCoS^3) and the Graduate School for Socio-Economic and Natural Sciences of the Environment (SENSE).

Joint PhD degree in Environmental Technology

UNIVERSITÉ — PARIS-EST

Docteur de l'Université Paris-Est

Spécialité: Science et Technique de l'Environnement

Dottore di Ricerca in Tecnologie Ambientali

UNESCO-IHE
Institute for Water Education

Degree of Doctor in Environmental Technology

Thése - Tesi di Dottorato - PhD thesis

Erika Jimena Espinosa-Ortiz

Bioreduction of selenite and tellurite by *Phanerochaete chrysosporium*

Defended on December 10th, 2015

In front of the PhD committee

Prof. Dr. Ir. Piet N.L. Lens	Promotor
Dr. Hab. Eric van Hullebusch	Co-Promotor
Dr. Hab. Giovanni Esposito	Co-Promotor
Dr. Eldon R. Rene	Mentor
Prof. Dr. François Farges	Examiner
Prof. Dr. Erkan Sahinkaya	Reviewer
Prof. Dr. Jonathan Lloyd	Reviewer

ETeCoS3

ERASMUS MUNDUS

Erasmus Joint Doctorate Programme in Environmental Technology for Contaminated Solids, Soils and Sediments (ETeCoS3)

CRC Press/Balkema is an imprint of the Taylor & Francis Group, an informa business

Published by:
CRC Press/Balkema
PO Box 11320, 2301 EH Leiden, The Netherlands
E-mail: Pub.NL@taylorandfrancis.com
www.crcpress.com – www.taylorandfrancis.com

ISBN 978-1-138-03004-6 (Taylor & Francis Group)

It always seems impossible,
until it is done.

Contents

Acknowledgments

Even though the following dissertation is a representation of an individual work, to me, it represents the summary of efforts of many people who offered their help, support and guidance. Without all of them I could never have reached the heights or explored the depths during this journey.

Firstly, I would like to address the Erasmus Mundus ETeCoS^3 program (FPA no.2010-0009) for financially supporting this research. I would also like to thank UNESCO-IHE, University of Paris-Est, University of Cassino and Southern Lazio, and the Center for Biofilm Engineering, for hosting me. I would also like to express my gratitude to Prof. Piet Lens for the continuous supervision of my PhD study and related research. Special thanks to my co-promotor, Dr. Eric van Hullebusch for his support, insightful comments and encouragement. Thanks Eric for your support and help, especially during the last months of my PhD. My sincere thanks also go to Prof. Robin Gerlach and Prof. Brent Peyton, who provided me the opportunity to join their team, and who gave me access to the laboratory and research facilities at the Center for Biofilm Engineering. I would also like to thank Prof. Giovanni Esposito, for his help with the administrative affairs related to the PhD program.

I thank Dr. Eldon Rene, not only for being my supervisor during the past three years, but also for his friendship. Thanks a lot Eldon for all the help you provided, your comments, your ideas, your scientific and moral support. I look up to you and the work you do. I would also like to express my gratitude to Prof. François Guyot for helping with the TEM analysis and enlighten me in the world of mineralogy. My appreciation also goes to Dr. Ellen Lauchnor and Dr. Yoan Pechaud, for their insights and help to better understand the complex world of the biofilms. I would also like to thank the master students I had the chance to work with, Manisha Shakya and Octavio García-Depraec. Thank you guys for all your hard work. My gratitude also goes to IHE-lab staff, Fred, Berend, Ferdi, Peter, Frank and Lyzette. Thank you guys for all your help in the lab. Thanks Peter for showing me how to cultivate my fun-guys for the very first time! Life in the lab was so much easier and fun with all of you around.

I would also like to thank Prof. Erkan Sahinkaya, Prof. Jonathan Lloyd and Prof. François Farges, jury members, for their critical comments and suggestions regarding my work in this dissertation. Your questioning towards my work had led me thinking more deeply about my research and had provided me with a new perspective.

Over the past years some colleagues had great contribution to my research. I would like to express my gratitude and admiration to my dear selenium-buddy and guru, Rohan Jain. Thanks a lot Rohan for all your great ideas and discussions. Sharing my professional and non-professional life with you for the past three years was a lot of fun. I look up to you my dear friend, and I hope one day I can have the same passion and knowledge as you do. Thanks also to Suthee, whose crazy way of thinking also pushed me to work harder and to be more creative to get the things I wanted to get. Also, thanks to Joy, the little selenium brother, for the discussions, collaborations and friendship.

This section could not be completed without acknowledging my friends, members of my secondary-adopted family that I have gathered in the past three years. Thanks a lot to *the girls*, amazing group of beautiful, smart and kind women that I had the chance to meet at UNESCO-IHE:

Angelica, Maribel, Fer, Jessi, Vero, Julliete and Pato. Thanks a lot for your support in the lowest moments of this journey, and for sharing the happy ones. My gratitude to my dear friends, Berend, Mohaned and Eduardo, whose support, love and friendship keep me smiling. Thanks to all the fellows that I had the chance to share some experiences with at UNESCO-IHE. Also my appreciation for the group of friends who helped me during my stay in the States, Luisa, Analy, Fede, Dayane, Rosa, Andrea and Chiachi. Thanks for helping me when in despaired, for cooking for me when I was sick and for driving me almost everywhere! Also, many thanks to the colleagues in Paris, Carolina, Ana, Anne, Alex, Douglas, Taam, Boris, Ania and Jules, for making the last days of the PhD bearable.

Finally, I would like to express my deepest gratitude to my family. Thanks to my mom, brother, aunts, uncles and cousins, for supporting me despite the distance and time. Thanks a lot mom for inspiring me every day. You gave me wings to fly, for that, I will always be grateful. It is ironic that it was the distance what brought us together, now you are not just my mom, but also my best friend. Last, but not least, I want to thank the person who made me who I am today, my granny. You are not longer here, but your teachings and love are always with me.

Summary

Selenium (Se) and tellurium (Te) are elements, they are part of the chalcogens (VI-A group of the periodic table) and share common properties. These metalloids are of commercial interest due to their physicochemical properties, and they have been used in a broad range of applications in advanced technologies. The water soluble oxyanions of these elements (*i.e.*, selenite, selenate, tellurite and tellurate) exhibit high toxicities, thus their release in the environment is of great concern. Different physicochemical methods have been developed for the removal of these metalloids, mainly for Se. However, these methods require specialized equipment, high costs and they are not ecofriendly. The biological treatment is a green alternative to remove Se and Te from polluted effluents. This remediation technology consists on the microbial reduction of Se and Te oxyanions in wastewater to their elemental forms (Se^0 and Te^0), which are less toxic, and when synthesized in the nano-size range, they can be of commercial value due to their enhanced properties. The use of fungi as potential Se- and Te-reducing organisms was demonstrated in this study. Response of the model white-rot fungus, *Phanerochaete chrysosporium*, to the presence of selenite and tellurite was evaluated, as well as their potential application in wastewater treatment and production of nanoparticles. The presence of Se and Te (10 mg L^{-1}) had a clear influence on the growth and morphology of the fungus. *P. chrysosporium* was found to be more sensitive to selenite. Synthesis of Se^0 and Te^0 nanoparticles entrapped in the fungal biomass was observed, as well as the formation of unique Se-Te nanocomposites when the fungus was cultivated concurrently in the presence of Se and Te. Potential use of fungal pellets for the removal of Se and Te from semi-acidic effluents (pH 4.5) was suggested. Moreover, the continuous removal of selenite (10 mg Se L^{-1} d^{-1}) in a fungal pelleted reactor was evaluated. The reactor showed to efficiently remove Se at steady-state conditions (~70%), and it demonstrated to be flexible and adaptable to different operational conditions. The reactor operated efficiently over a period of 35 days. Good settleability of the fungal pellets facilitated the separation of the Se from the treated effluent. The use of elemental Se immobilized fungal pellets as novel biosorbent material was also explored. This hybrid sorbent was promising for the removal of zinc from semi-acidic effluents. The presence of Se in the fungal biomass enhanced the sorption efficiency of zinc, compared to Se-free fungal pellets. Most of the research conducted in this study was focused on the use of fungal pellets. However, the response of the fungus to selenite in a different kind of growth was also evaluated. Microsensors and confocal imaging were used to evaluate the effects of Se on fungal biofilms. Regardless of the kind of fungal growth, *P. chrysosporium* seems to follow a similar selenite reduction mechanism, leading to the formation of Se^0. Architecture of the biofilm and oxygen activity were influenced by the presence of Se.

Résumé

Le sélénium et le tellurium partagent des propriétés chimiques communes et appartiennent à la colonne des éléments chalcogènes de la classification périodique des éléments. Ces métalloïdes ont des propriétés physico-chimiques remarquables et ils ont été utilisés dans un grand nombre d'applications dans le domaine des hautes technologies (électronique, semi-conducteurs, alliages). Ces éléments, qui se retrouvent généralement sous formes d'oxyanions, sont extrêmement solubles dans l'eau et présentent une forte toxicité. Leur libération dans l'environnement représente donc un enjeu capital. Différentes méthodes physico-chimiques ont été développées pour la récupération de ces metalloïdes, en particulier pour le sélénium. Néanmoins, ces méthodes requièrent un équipement lourd et coûteux et ne sont pas très recommandables sur le plan écologique. Le traitement biologique est donc une bonne alternative pour la récupération de Se et de Te provenant des effluents pollués. Cette approche réside dans la bioréduction des différents oxyanions sous formes métalliques. Ceux-ci sont moins toxiques et d'intérêts commerciales notables surtout lorsqu'ils se présentent sous forme nanométrique. L'utilisation de micro-champignons comme microorganismes catalyseur de la réduction de Se et de Te a été démontrée dans cette étude. La réactivité du micro-champignon responsable de la pourriture blanche, *Phanerochaete chrysosporium* en présence de sélénite et de tellurite a été évaluée, ainsi que son application potentielle pour le traitement des eaux contaminées et la production de nanoparticules. La présence de Se et de Te (10 mg L^{-1}) a une influence importante sur la croissance et la morphologie du micro-champignon. Il s'avère que *P. chrysosporium* est très sensible à la présence de sélénites. La synthèse de Se° et de Te° sous forme de nanoparticules piégées dans la biomasse fongique a été observée, ainsi que la formation de nano-composites Se-Te lorsque le champignon était cultivé simultanément en présence des deux métalloïdes. L'usage potentiel de biofilm fongiques pour le traitement des effluents semi-acides (pH 4.5) contenant du Se et du Te a été suggéré. De plus, le traitement en mode continu de sélénite (10 mg L^{-1} d^{-1}) dans un réacteur à biofilm fongique granulaire a été évalué. Le réacteur a montré un rendement d'élimination du sélénium en régime permanent de 70% pour differentes conditions opératoires. Celui-ci s'est montré efficace pendant une période supérieure à 35 jours. La bonne sédimentation du biofilm granulaire facilite la séparation du sélénium de l'effluent traité. L'utilisation du biofilm granulaire contenant du sélénium élémentaire comme bio-sorbant a également été étudiée. Cet adsorbant hybride s'est montré prometteur pour l'immobilisation du zinc présent dans les effluents semi-acides. La plupart des recherches effectuées se sont focalisées sur l'utilisation des biofilms granulaires. Toutefois, la croissance du micro-champignon suite à l'exposition à des concentrations différentes de sélénites a également été étudiée. Des micro-électrodes à oxygène et un microscope confocal à balayage laser ont été utilisées pour évaluer l'effet du sélénium sur la structure des biofilms fongiques. Quel que soit le mode de croissance de *P. chrysosporium*, le mécanisme de réduction du sélénite semble être toujours le même tout en menant à la formation de sélénium élémentaire. Cependant, l'architecture des biofilms et les profils en oxygène sont influencées par la présence de sélénium.

Sommario

Il selenio e il tellurio sono elementi peculiari. Essi fanno farte de calcogeni (VI-A gruppo della tavola periodica) e ne condividono le proprietà comuni. Tali metalloidi devono il proprio interesse commerciale alle proprietà fisico-chimiche, e sono stati utilizzati in un ampio campo di applicazioni e tecnologie avanzate. Gli osso-anioni idrosolubili di talielementi (*i.e*, selenito, selenato, tellurito e tellurato) esibiscono alta tossicità, e dunque il loro rilascio in natura è di grande interesse. Metodi chimico-fisici differenti sono stati sviluppati per la rimozione di questi metalloidi e in particolar modo per il selenio. Ad ogni modo, tali metodi richiedono attrezzature specifiche ed alti costi, oltre a non risultare ecologici. Alternativa "verde" per la rimozione di Se e Te da effluenti contaminati è il trattamento biologico. Questa tecnologia di trattamento consiste nella riduzione, ad opera di microorganismi, degli osso-anioni di Se e Te alle loro forme elementari e meno tossiche (Se^0 e Te^0), che sintetizzate nella gamma delle nano-dimensioni, possono assumere valore commerciale in virtù delle loro intensificate proprietà. In questo studio, è stato valutato l'uso di funghi, come potenziale organismo riducente di Se e Te. È stata quindi studiata la risposta di un fungo modello della famiglia"white-rot", ovvero *Phanerochaete chrysosporium*, in presenza di selenito e tellurito, così come la sua potenziale applicazione nel trattamento di acque reflue e nella produzione di nanoparticelle. La presenza di Se e Te (10 mg L^{-1}) ha avuto una chiara influenza sulla crescita e sulla morfologia del fungo. *P.chrysosporium* si è rivelata essere più sensibile al selenito. È stata osservata, infatti, quando la coltivazione del fungo è stata condotta in presenza di Se e Te, la sintesi di nanoparticelle di Se^0 e Te^0 intrappolate nella biomassa, così come la formazione di nano-composti unici di Se-Te. Tale risultato ha suggerito quindi il potenziale uso di pellet fungino per la rimozione di Se e Te da effluenti semi-acidi (pH 4,5). Inoltre, è stata valutata la rimozione del selenito (10 mg L^{-1} d^{-1}) con pellet fungini in un reattore in continuo. Il reattore si è dimostrato capace di rimuovere efficacemente il selenio in condizioni stazionarie (intorno al 70%), e ha dimostrato di essere flessibile e adattabile a differenti condizioni operative. Il reattore ha lavorato efficientemente per un periodo di oltre 35 giorni. La buona capacità di sedimentazione dei pellet fungini ha facilitato la separazione del Selenio dall'effluente trattato. È stato esaminato, inoltre, l'uso di pellet fungino di selenio elementare immobilizzato come un nuovo materiale bio-assorbente. Questo assorbente ibrido è risultato promettente per la rimozione di zinco da effluenti semi-acidi. La presenza di Selenio nella biomassa fungina ha infatti aumentato l'assorbimento dello zinco, comparato all'assorbimento prodotto dall'utilizzo di pellet fungini privi di Se. La maggior parte delle ricerche condotte in questo studio si sono concentrate sull'uso di pellet fungino. Tuttavia, è stata anche esaminata la risposta del fungo al selenito in differenti condizioni di crescita. Sono stati utilizzati micro-sensori e immagini confocali per valutare gli effetti del selenio sul biofilm fungino. Qualunque sia il tipo di crescita fungina, il *P. chrysosporium* sembra seguire lo stesso meccanismo di riduzione del selenito, portando alla formazione di Se^0. L'architettura del biofilm e l'attività aerobica sono state entrambe influenzate dalla presenza di selenio.

Samenvatting

Seleen (Se) en telluur (Te) zijn bijzondere elementen, ze maken deel uit van de zuurstofgroep (VI-A groep in het periodieke stelsel). Ze delen een aantal gemeenschappelijke eigenschappen. Door hun fysische eigenschappen zijn deze metalloïden al breed toegepast in geavanceerde technologieën en dat maakt ze interessant voor commerciële doeleinden. De wateroplosbare anionen van deze elementen (zoals Se(IV), Se(VI), Te(IV) en Te(VI)) zijn zeer giftig en het is daarom een groot probleem als deze elementen in het milieu terecht komen. Er zijn, met name voor Se, verschillende fysische methodes ontwikkeld om deze metalloiden uit afvalwater te verwijderen. Deze methodes vergen speciale apparatuur, er zijn hoge kosten aan verbonden en ze zijn niet milieuvriendelijk. Biologische behandeling is een groen alternatief waarmee Se en Te verwijderd kunnen worden uit vervuild afvalwater. Deze saneringstechnologie maakt gebruik van een microbiële reductie om de Se en Te anionen in afval water om te vormen naar hun minder giftige elementaire vorm (Se^0 en Te^0). Wanneer deze elementaire vorm als nanodeeltjes gesynthetiseerd worden hebben ze door hun eigenschappen ook nog commerciële waarde. In deze studie wordt het gebruik van schimmels als mogelijke Se- en Te- reducerende organismen aangetoond. De reactie van *Phanerochaete chrysosporium*, op de aanwezigheid van seleniet en telluriet, hun mogelijke toepassing in de behandeling van afvalwater en de productie van nanodeeltjes is geëvalueerd. De aanwezigheid van Se en Te (10 mg L^{-1}) had een duidelijke invloed op de groei en morfologie van de schimmels. Het blijkt dat *P. chrysosporium* gevoeliger is voor Se(IV). Ook de synthese van Se^0 en Te^0 nanodeeltjes gevangen in biomassa is waargenomen. Wanneer de schimmel werd gekweekt in aanwezigheid van Se en Te is de vorming van het unieke Se-Te nanocomposieten geconstateerd. In dit onderzoek wordt het gebruik van schimmel pellets voor het verwijderen van Se en Te in semi-zuur afvalwater (pH 4.5) voorgesteld. Verder is het continue verwijderen van Se(IV) (10 mg L^{-1} d^{-1}) in een schimmel pellet reactor geëvalueerd. Bij constante condities bleek de reactor efficiënt in het verwijderen van seleen (~70%). Ook is de reactor aantoonbaar flexibel en aanpasbaar voor verschillende operationele omstandigheden. De reactor opereerde efficiënt over een periode van 35 dagen. Goede bezinking van de schimmel pellets vergemakkelijkt de afscheiding van seleen uit het te behandelen afvalwater. Het gebruik van elementair seleen geïmmobiliseerde schimmel pellets als nieuw biosorbent materiaal werd ook onderzocht. Dit hybride sorbens is veelbelovend voor het verwijderen van zink uit semi-zuur afvalwater. In vergelijking met schimmel pellets zonder Se verbeterde de aanwezigheid van seleen de sorptie efficiëntie van zink. Het merendeel van het onderzoek in deze studie was gericht op het gebruik van schimmel pellets, maar de reactie van de schimmels op Se(IV) in een andersoortige groei is ook geëvalueerd. Microsensoren en confocale beeldvorming werden gebruikt om de effecten van seleen op schimmel biofilms te evalueren. Ongeacht de groeivorm van de schimmel volgt *P. chrysosporium* een vergelijkbaar Se(IV) reductie mechanisme, dat leidt tot de vorming van Se^0. Zowel de architectuur en zuurstof activiteit van de biofilm werden beïnvloed door de aanwezigheid van seleen.

CHAPTER 1

GENERAL INTRODUCTION

1.1 Background

Selenium (Se) and tellurium (Te) metalloids belonging to the chalcogen's group (VI-A group of the periodic table) are of commercial interest due to their physicochemical properties. The use of Se and Te in advanced technologies includes a broad range of applications, *e.g.,* alloying agents, semiconductors, electronics, solar cells and other functional materials. The interest towards these metalloids has increased over the last decades, particularly as nanoparticles, due to their enhanced properties compared to their bulk materials. In nature, Se and Te can be found in different oxidation states, including +VI, +IV, 0 and -II. The water soluble oxyanions of both elements (+VI and +IV) are highly bioavailable and toxic, thus their release in the environment is of great concern. As an essential trace element, the World Health Organization (2011) has established 40 ug L^{-1} as the maximum daily intake of Se; whereas, the Environmental Protection Agency (2015) has recommended 5 ug L^{-1} of Se as the water quality criterion for the protection of aquatic life. Te is not an essential trace element, but its toxicity towards most microorganisms has been documented; for some bacteria even at concentrations as low as 1 mg L^{-1} (Taylor, 1999; Díaz-Vásquez et al., 2015; Reinoso et al., 2012).

Polluted effluents containing either Se or Te are usually related to agricultural and industrial activities, including mining, refinery, coal combustion and electronic industries, among others (Lenz and Lens, 2009; Perkins, 2011; Biver et al., 2015). The application of biological agents is an ecofriendly and cost-effective alternative to the traditional physicochemical treatments (Henze et al., 2008) to remove Se oxyanions from polluted effluents. Different microorganisms have been demonstrated to be capable of reducing the toxic Se and Te oxyanions (selenate -SeO_4^{2-}, selenite -SeO_3^{2-}, terullate -TeO_4^{2-} and tellurite -TeO_3^{2-}) to their elemental forms (Se^0 and Te^0), which are more stable and less toxic (Chasteen et al., 2009; Winkel et al., 2012). Moreover, the biological reduction of Se and Te oxyanions has led to the synthesis of Se^0 and Te^0 particles in the nano-size range.

Most typically, bacteria are the preferred organisms for the treatment of wastewaters. However, the use of fungi is also promising. Fungi possess the unique ability to produce large amounts of enzymes and reductive proteins, which makes them highly valuable in different biotechnological applications. These heterotrophic organisms have been used in the food and chemical industry, medicine and bioremediation (Archer 2000; Khan et al., 2014; Espinosa-Ortiz et al., 2016). As alternative agents in the biological treatment, fungi have been found to be capable to efficiently remove several kinds of different organic and inorganic pollutants (Espinosa-Ortiz et al., 2016). Fungi have also been demonstrated to be Se- and Te-reducing organisms, capable of biomineralizing Se^0 and Te^0 (Gharieb et al., 1995; Gharieb et al., 1999). Nevertheless, the information on the influence of these metalloids on fungi in relation to their environment, competition or survival is scarce.

1.2 Problem statement

Se and Te are rare trace elements irregularly distributed in earth's crust, usually recovered as by-products from metal bearing ores. The consumption of these elements has increased in recent years, and it is projected to continue growing. In fact, Se and Te have been identified among the 14 elements for which the European demand will require >1% of the current world supply per annum between 2020 and 2030 (Moss et al., 2011). The scarcity of these metalloids makes imperative to explore for new sources or methods to recover them.

Biomineralization of Se and Te composites is a promising alternative to recover these elements from polluted effluents containing their oxyanions. The development of a technology that not only allows removing Se and Te pollutants from wastewaters, but also allows producing highly valuable resources in the form of Se and Te nano-biocomposites is desirable. Most of the research on the development of such technology has been limited to the use of bacteria. Assessing the tolerance and evaluating the potential use of other organisms against significant concentrations of Se and Te would enable to screen suitable species for Se and Te biotransformation and biorecovery from polluted effluents. The use of fungi as Se- and Te-reducing organisms is promising. Interaction between Se, Te and fungi has been found in the literature since the early 1900s (Stower and Hopkins, 1927). However, the information regarding fungal resistance and response to Se and Te oxyanions is scarce. Although several fungal strains possess the ability to reduce these metalloids, their use for removal of Se and Te polluted effluents has not yet been exploited. Moreover, the characterization of Te and Se biocomposites synthesized by fungi has barely been reported.

Even though fungi have shown to be excellent degradative agents with potential for bioremediation purposes, little attention has been given to the use of fungal bioreactors, particularly using fungal pellets. The progress of large scale and cost-effective application of fungi to continuous treatment of liquid effluent has been delayed by the lack of suitable bioreactor systems. Different fungal bioreactors have been proposed to overcome this limitation, being the most important ones: airlift (Ryan et al., 2005) and pulsed system fluidized-bed reactors (Lema et al., 2001). Even though efforts have been made to continue developing fungal bioreactors, this field is still considered to be in its primitive stage. Most of the applications are focused on treating organic pollutants (*e.g.*, dyes, phenols). So far, there are no publications on the use of fungal bioreactors to treat Se pollutant compounds; hence there are no propositions to develop a fungal bioreactor system that allows not only treating Se contaminated streams, but also obtaining a profitable product such as Se^0 nanoparticles.

1.3 Research objectives

This thesis aims to provide a better understanding of the fungal interactions with the toxic Se and Te oxyanions, selenite (SeO_3^{2-}) and tellurite (TeO_3^{2-}), in order to explore their potential application and relevance in wastewater treatment and synthesis of nanoparticles. This would help to enlarge the scope of organisms used for Se and Te bioremediation and biomineralization. The specific objectives of this research were:

1) To investigate the response of fungal pellets of *P. chrysosporium* to the presence of SeO_3^{2-} and TeO_3^{2-}, in terms of growth, ability of substrate consumption and morphology.
2) To determine the potential of *P. chrysosporium*, as a Se^0, Te^0 and Se-Te nanofabric.
3) To determine the feasibility of fungal pelleted systems for the continuous removal of SeO_3^{2-} by assessing the performance of an up-flow fungal pelleted reactor.

4) To explore the possibility to use Se^0 nanoparticles immobilized fungal pellets as novel sorbent material for the removal of zinc from semi-acidic polluted effluents.
5) To study the response of *P. chrysosporium* to SeO_3^{2-} in a different type of fungal growth, as biofilm, by assessing the temporal and spatial effects in the oxygen activity of the biofilm with the help of dissolved oxygen microsensors.

1.4 Structure of the thesis

This thesis is comprised by eight chapters. Figure 1.1 shows the flow diagram of the structure of this dissertation. *Chapter 1* provides a general overview of the content of this dissertation, including the background, problem statement, research objectives and the description of the structure of the thesis. *Chapter 2* corresponds to the literature review concerning the formation of fungal pellets, their use in bioreactors and their applications for the removal of organic and inorganic pollutants from water. This chapter has been published as Espinosa-Ortiz et al. (2016a). *Chapter 3* describes the effects of Se oxyanions on *P. chrysosporium* pellets. Influence of selenite in the growth, substrate consumption and morphology of the pellets was determined. The effect of different operational parameters in the Se-fungus interactions and Se removal were also determined. Mycosynthesis of Se^0 nanoparticles is also demonstrated in this chapter, which was published as Espinosa-Ortiz et al. (2015a). *Chapter 4* explores the potential use of fungal pellets for the removal of selenite from semi-acidic effluents (pH 4.5) in a continuous up-flow bioreactor. Different operational conditions, Se loading rates, Se spikes and strategies to control the fungal growth were examined. This chapter was published as Espinosa-Ortiz et al. (2015b). *Chapter 5* investigates the use of Se^0 nanoparticles immobilized in fungal pellets as novel sorbent materials for the removal of zinc from mild acidic effluents (pH 4.5). *Chapter 6* describes the influence of selenite on fungal biofilms through the use of oxygen microsensors and imaging with confocal laser scanning microscopy, to determine the changes on the oxygen consumption within the biofilm and the influence on the biofilm architecture. This chapter was published as Espinosa-Ortiz et al. (2016b). *Chapter 7* includes the information regarding the effects of tellurite on fungal pellets. The fungal response to the concurrent presence of selenite and tellurite was also investigated. The fungal biomineralization of nSe^0, nTe^0 and nSe-Te biocomposites is reported. *Chapter 8* summarizes the information gathered in this dissertation and provides the insights on the potential use of fungal pellets for the treatment and biomineralization of Se and Te oxyanions.

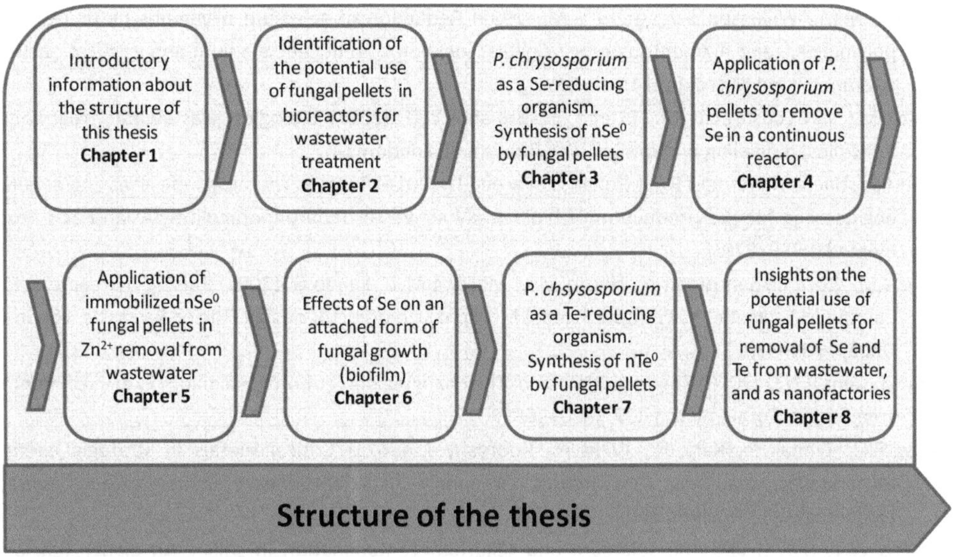

Figure 1.1 Flow diagram of the structure of this dissertation.

1.5 References

Archer D.B. (2000) Filamentous fungi as microbial cell factories for food use. Curr Opin Biotechnol 11:478–483.

Biver M., Quentel F., Filella M. (2015) Direct determination of tellurium and its redox speciation at the low nanogram level in natural waters by catalytic cathodic stripping voltammetry. Talanta.

Chasteen T.G., Fuentes D.E., Tantaleán J.C., Vásquez C.C. (2009) Tellurite: history, oxidative stress, and molecular mechanisms of resistance. FEMS Microbiol Rev 33:820–832.

Díaz-Vásquez W.A., Abarca-Lagunas M.J., Cornejo F.A., Pinto C., Arenas F., Vásquez C.C. (2015) Tellurite-mediated damage to the *Escherichia coli* NDH-dehydrogenases and terminal oxidases in aerobic conditions. Archives Biochem Biophys 566:67–75.

Espinosa-Ortiz E.J., Gonzalez-Gil G., Saikaly P.E., van Hullebusch E.D., Lens P.N.L. (2015a) Effects of selenium oxyanions on the white-rot fungus *Phanerochaete chrysosporium*. Appl Microbiol Biotechnol 99(5):2405–2418.

Espinosa-Ortiz E.J., Rene E.R., van Hullebusch E.D., Lens P.N.L. (2015b) Removal of selenite from wastewater in a *Phanerochaete chrysosporium* pellet based fungal bioreactor. Int Biodeterior Biodegradation 102:361–369.

Espinosa-Ortiz E.J., Rene E.R., Pakshirajan K., van Hullebusch E.D., Lens P.N.L. (2016a) Fungal pelleted reactors in wastewater treatment: applications and perspectives. Chem Eng J 283:553–571.

Espinosa-Ortiz E.J., Pechaud Y., Lauchnor E., Rene E.R., Gerlach R., Peyton B., van Hullebusch E.D., Lens P.N.L. (2016b) Effect of selenite on the morphology and respiratory activity of *Phanerochaete chrysosporium* biofilms. Bioresour Technol. doi: 10.1016/j.biortech.2016.02.074.

Gharieb M.M., Kierans M., Gadd G.M. (1999) Transformation and tolerance of tellurite by filamentous fungi: accumulation, reduction, and volatilization. Mycol Res 103:299–305.

Gharieb M.M., Wilkinson S.C., Gadd G.M. (1995) Reduction of selenium oxyanions by unicellular, polymorphic and filamentous fungi: Cellular location of reduced selenium and implications for tolerance. J Ind Microbiol 14:300–311.

Henze M., van Loosdrecht M., Ekama G., Brdianovic D. (2008) Biological wastewater treatment: Principles, modelling and design. IWA Publishing, London, UK.

Khan A.A., Bacha N., Ahmad B., Lutfullah G., Farooq U. (2014) Fungi as chemical industries and genetic engineeriing for the production of biologically active secondary metabolites. Asian Pac J Trop Biomed 4:859–870.

Lema J.M., Roca E., Sanroman A., Nunez M.J., Moreira M.T., Feijoo G. (2001) Pulsing bioreactors. In: Cabral J.M.S., Mota M., Tramper J. (eds) Multiphase bioreactor design. Taylor & Francis, London, UK, pp. 309–329.

Lenz M., Lens P.N.L. (2009) The essential toxin: The changing perception of selenium in environmental sciences. Sci Total Environ 407:3620–3633.

Moss R.L., Tzimas E., Kara H., Willis P., Kooroshy J. (2011) Critical metals in strategic energy technologies, assessing rare metals as supply-chain bottlenecks in low-carbon energy technologies. EUR 24884 EN.

Perkins W.T. (2011) Extreme selenium and tellurium contamination in soils - An eighty year-old industrial legacy surrounding a Ni refinery in the Swansea Valley. Sci Total Environ 412-413:162–169.

Reinoso C., Auger C., Appanna V.D., Vásquez C.C. (2012) Tellurite-exposed Escherichia coli exhibits increased intracellular α-ketoglutarate. Biochem Biophys Resear Comm 421(4):721–726.

Ryan D.R., Leukes W.D., Burton S.G. (2005) Fungal bioremediation of phenolic wastewaters in an airlift reactor. Am Chem Soc Am Inst Chem Eng 21:1068–1074.

Stower N.M., Hopkins B.S. (1927) Fungicidal and bactericidal action of selenium and tellurium compounds. Ind Eng Chem 19:510–513.

Taylor D.E. (1999) Bacterial tellurite resistance. Trends Microbiol 7:111–115.

US EPA. (2015) Draft aquatic life ambient water quality criterion for selenium (freshwater).

WHO. (2011) Selenium in drinking water. Background document for development of WHO *Guidelines for Drinking-water Quality.*

CHAPTER 2

LITERATURE REVIEW

A modified version of this chapter was published as:

E.J. Espinosa-Ortiz, E.R. Rene, K. Pakshirajan, E.D. van Hullebusch, P.N.L. Lens. (2016) Fungal pelleted reactors in wastewater treatment: Applications and perspectives. Chemical Engineering Journal. 283:553–571.

Abstract

The use of fungal species to remove organic and inorganic pollutants from wastewater has shown to be a good alternative to traditional wastewater treatment technologies. Fungal pellets are well settling aggregates formed by self-immobilization. Their use in bioreactors is promising as it avoids the practical and technical difficulties usually encountered with dispersed mycelium. This review presents the mechanisms involved in the formation and growth of fungal pellets as well as the different factors that influence the stability of the pellets. The various types of fungal pelleted bioreactors that are used for wastewater treatment, their configuration, design and performance are reviewed. A summary of the different organic and inorganic pollutants that have been treated using fungal pelleted reactors, from dyes to emergent pollutants such as pharmaceuticals, are discussed from an application view-point. The operational issues such as bacterial contamination and longevity of this bioprocess under non-sterile conditions, as well as the reuse of fungal pellets are also encompassed in this review.

Keywords: Fungal pellets, bioreactors, wastewater treatment, pellet stability, non-sterile conditions, reuse potential

2.1 Introduction

The major driving forces among industries to adopt biological systems for wastewater treatment is their low cost, energy saving efficiency and the production of valuable end-products that can be used for energy production or used as fertilizers (Zhang et al., 2011). Bioreactors are commonly employed biotechnologies for wastewater treatment. Most commonly, bacteria are used in bioreactors for the treatment of wastewater, whereas the use of fungi has received much less attention. Fungal bioreactors are advantageous due to the rich source of degrading enzymes produced by fungi as well as their ability to withstand harsh conditions, especially fluctuating pollutant loads, low pH and tolerance to low nutrient concentrations (Zhang et al., 1999; Jeyakumar et al., 2013; Kennes et al., 2004; Grimm et al., 2005).

Fungi are heterotrophic, eukaryotic and achlorophyllous organisms, which have the ability to secrete large amounts of enzymes and reductive proteins. Such unique properties make fungi attractive to produce different metabolites such as citric acid (Papagianni, 2007), lovastatine (Casas-López et al., 2003), lactic acid (Zhang et al., 2007), carotene (Nanou et al., 2012), and enzymes such as protease and laccase (Cabaleiro et al., 2002; Kandelbauer et al., 2004). Different fungal strains have shown their ability to degrade a wide range of environmental pollutants, from dyes to pharmaceutical compounds, heavy metals, trace organic contaminants and endocrine disrupting contaminants.

The use of fungal pellets in bioreactors has attracted the attention of environmental bioengineers owing to their potential operation in a continuous regime as well as their capacity to reduce the operational difficulties caused by fungal dispersed mycelium. Inferior mixing and oxygen supply, foaming and fungal growth on the bioreactor walls, agitators and baffles are identified operational problems. Most of the research on the application of fungal pellets for wastewater treatment has so far focused mainly on assessing the degradation capability of fungal pellets in small-scale, sterile batch tests.

This review overviews the advantages and limitations of using pelleted growth over dispersed mycelial growth in fungi-inoculated wastewater treatment reactors. The operational parameters to be considered for the design of fungal pelleted reactors are discussed briefly. The challenges of using fungal pelleted bioreactors for environmental applications and recent trends to overcome their operational challenges are also discussed.

2.2 Fungal pellets

2.2.1 Fungal pellets: formation and growth

Fungal pellets are spherical, ellipsoidal or oval masses of intertwined hyphae with a size usually in the range of several hundred micrometers to several millimeters (Braun and Vecht-Lifsitz, 1991; Nielsen and Villadsen, 1994; Cuie et al., 1998). Fungal pellets usually appear as a core of densely packed hyphae, surrounded to a large extent by a more annular dispersed or "hairy" region that contains the radially growing portion of the hyphae (Domingues et al., 2000). Four different regions can be recognized in a fungal pellet (Wittier et al., 1986): (i) the first region corresponds to a compact core region, which presents a semi-anaerobic environment with very little amount of viable hyphae, (ii) the second region corresponds to the layer surrounding the central region, the hyphae present an irregular wall structure, (iii) a third region appears in hollow pellets, with hyphae showing clear signs of autolysis, and (iv) the fourth region corresponding to the external hairy zone, the hyphae in this region are viable and metabolically more active than in the other regions (Figure 2.1).

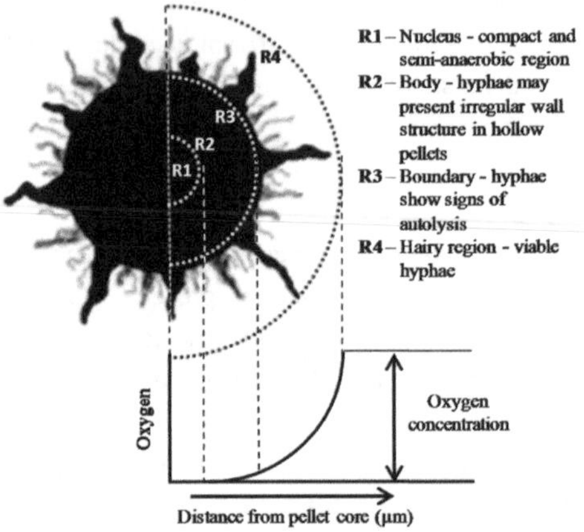

Figure 2.1 Regions of a fungal pellet and oxygen gradient within the fungal pellet.

Three main phases have been identified for the formation and growth of filamentous fungi in submerged cultures (Figure 2.2). The formation of pellets starts with the swelling and germinating phase of spores, followed by the hyphal growth and branching (Phase 1: micro-morphological growth). This process depends on the physicochemical properties of the spores and hyphae, as well as on the cultivation conditions such as the pH and salinity of the growth medium and its rheological behavior (Grimm et al., 2005). Pellets can be formed from either a single spore (non-coagulative), aggregates of spores (coagulative) or agglomerated hyphae (Nielsen, 1996). Once the branching of hyphae has initiated, the hyphal growth continues to Phase 2 (macro-morphological growth), accompanied by hyphae to hyphae interactions, leading to the formation of the pellets (Metz and Kossen, 1977). The decay phase (Phase 3: fungal cell autolysis) begins with the erosion of the pellets and finalizes with their breaking up. The fragmented hyphae pieces may then serve as new centers for pelletization (Papagianni, 2004).

Fungal pellets can be efficiently used as immobilized cell systems due to their morphological characteristics, which facilitate the cross-linking or cell entrapment without the use of chemical agents. Previous studies have reported the formation of biomass pellets during the growth of different filamentous fungal cultures, including *Cladosporium cladosporoides, Aspergillus nidulans, Rhizopus oryzae, Anthracophyllum discolor, Trichoderma reesei* and *Pleurotus ostreatus* (Kim and Song, 2009; Liao et al., 2007; Rubilar et al., 2009; Yu et al., 2012).

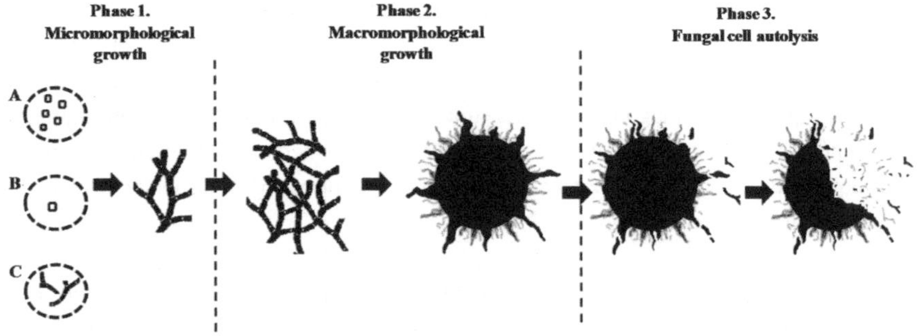

Figure 2.2 Filamentous fungal growth in submerged cultures; Phase 1-Micromorphological growth, swelling and germinating phase of spores, hyphal growth and branching. Pellets can be originated from: A) coagulating spores, B) a single spore, or C) agglomerated hyphae; Phase 2-Macromorphological growth, hyphal networking and formation of pellets; Phase 3-Fungal cell autolysis, erosion and breaking up of the pellets.

2.2.2 Factors influencing the characteristics and formation of pellets

The growth and morphology of filamentous fungi in submerged cultures depend on different operational conditions and factors, namely, (i) inoculation, (ii) medium composition, and (iii) cultivation conditions. The inoculation dependent factors include the type of fungal strain used, the quality and amount of inoculum, and the inoculation process. Not all fungal species or strains produce pellets under the same conditions. The amount of inoculum used dictates the growth and morphology in submerged cultures. It has been suggested that using high concentrations of fungal spores might lead to the formation of dispersed mycelia, whereas at a low inoculum level (usually <10^8 spores mL^{-1}), the formation of pellets is induced (Papagianni and Moo-Young, 2002). The production of fungal pellets is traditionally done with a spore solution. However, the successful production of pellets using mycelium fragments has also been reported (Xin et al., 2012). The inoculum size also impacts the pelleted growth (Liao et al., 2007), and an inversely proportional relationship was reported between the amount of inoculum used and pellet size (Sharma and Padwal-Desai, 1985). Furthermore, the inoculum age has been shown to influence the final pellet diameter (Colin et al., 2013).

The medium composition, in particular the carbon source, nutrients and some additives in the fungal broth are key factors for the pelletization process. The influence of different carbon sources (*e.g.*, sucrose, glucose, fructose and xylose) on the growth morphology of different fungal strains has been widely reported (Cho et al., 2002; Javanbakht et al., 2011; Jonsbu et al., 2002; Rodríguez-Couto, 2012). Réczey et al. (1992) reported that various strains of *Aspergillus niger* grew as free mycelia when lactose was used as a carbon source. However, pellets were formed when this fungal strain was grown with glucose as the substrate. The use of additives such as polymers, surfactants or chelators is a successful strategy to influence the morphology of filamentous fungi, either by preventing or enhancing the formation of pellets according to the desired objective. The addition of polymers (*e.g.*, carbopol, polyacrylic acid) might reduce the capacity of spores to aggregate, thus, limiting pellet formation (Saraswathy and Hallberg, 2005). In contrast, the use of surfactants enables naturally filamentous organisms to form pellets. This is the case for the naturally filamentous *Trichoderma reesei*, which in the presence of Triton X-100 and the biosurfactant rhamnolipid was able to grow as fungal pellets (Callow and Ju, 2012). The addition of certain toxic compounds has also shown to influence fungal morphology. In a previous study, the presence of selenite, a toxic soluble oxyanion of selenium, promoted the formation of more compact and smother pellets of *Phanerochaete chrysosporium* (Espinosa-Ortiz et al., 2015a). Osmolality not only affects fungal morphology, but also its productivity (Wucherpfennig et al., 2011). Fungal pellets were found to be more elongated and showed an increase in their surface area as the osmolality of the medium was raised. Moreover, higher sodium chloride concentrations (>4.2 osmole kg^{-1}) led to the formation of dispersed mycelia, without any clumps or pellets (Wucherpfennig et al., 2011).

The morphology of the pellets is further influenced by the hydrodynamic conditions prevailing within a bioreactor. García-Soto et al. (2006) showed the effect of different hydrodynamic regimes (*i.e.,* laminar, transitional and turbulent), according to the friction factor in the bioreactor based on the Reynolds number (Bird et al., 2002) on the morphology of fungal pellets of *R. nigricans* in a bubble column reactor. Pellets were conserved in the laminar regime, *i.e.*, mostly homogeneous flow inside the column, and the fungal pellets did not show any signs of significant erosion. However, for bioreactor operation during transitional and turbulent regimes, destruction/fragmentation of the initial pellets was observed. This promoted the presence of small compact nuclei. The authors reported that intense mixing zones usually found in the transitional and turbulent regimes forced the mycelia to stretch and roll into themselves, leading to more compact and ovoid forms (García-Soto et al., 2006).

The intensity of agitation affects the degree of compactness or fluffiness of pellets: small and dense pellets formed under intense agitation (>600 rpm) and vice versa (Rodríguez-Porcel et al., 2005). The surface properties of fungi and the strong short range cohesive interactions facilitate hyphae-hyphae linkages, while weak long range interactions form freely dispersed mycelial cells. The mode of aeration provided to the fungal system also plays an important role in determining the morphology of the pellets. Continuous air flow favors excessive fungal growth (Moreira et al., 1996), whereas the use of oxygen pulses controls the size of the pellets and its morphology, avoiding excessive biomass growth (Rodarte-Morales et al., 2012). Moderate to high levels of O_2 improve the growth and morphological features of fungi, while in some cases, antagonistic effects are attributed to oxidative stress at higher O_2 concentrations (Bai et al., 2003). Under oxygen rich conditions, the morphology of fungi changes from dispersed filamentous to aggregate form, as clumps (Kreiner et al., 2003) or pellets (Higashiyama et al., 1999), by developing highly branched hyphae as a self-protection strategy against oxidative stress. The size of the pellets is also influenced by high O_2 concentrations. Nanou et al. (2011) demonstrated that oxidative stress leads to the formation of smaller pellets during carotene production in a bubble column reactor by *Blakeslea trispora*.

2.2.3 Fungal pellets in bioreactors

The type of fungal growth preferred for a particular process depends on the application. In suspended growth bioreactors, three forms of fungal morphologies can be present: suspended mycelia (freely dispersed filaments), clumps (aggregated but still dispersed) and pellets (denser, spherical aggregated forms) (Paul and Thomas, 1998; Pazouki and Panda, 2000). Table 2.1 summarizes the main advantages and limitations of fungal pelleted growth in bioreactor applications. For wastewater treatment applications, the growth of dispersed mycelia is not desirable due to the operational difficulties that filamentous fungi cause. The presence of dispersed hyphae-like elements in the cultivation broth increases its viscosity, leading to a change in the rheological behavior (non-

Newtonian fluid), which in turn may cause insufficient mixing and thus oxygen and nutrient deficiency (Metz et al., 1979). Excessive mycelial growth can cause fungal growth on the walls of the reactors, agitators, baffles, sampling and nutrient addition lines.

Advantages of immobilized fungal cells include the easiness of separating cells from the medium, which makes the reproducibility of batch cultivation feasible and allows continuous product recovery. The rheology of the fungal culture is usually improved by the pelleted biomass, presenting a lower apparent viscosity and enhancing the oxygen and mass transfer into the biomass pellets with a reduced energy demand for aeration and agitation (Thongchul and Yang, 2003). During continuous or fed-batch operation, higher cell loading and volumetric productivities have been reported, since fungal cells can withstand better shear forces than bacterial cells. Some disadvantages of working with biomass pellets include the alterations in fungal cell physiology and kinetics due to mass transfer and oxygen limitations, leading to autolysis in the interior of large pellets (El-Enshasy, 2007). Wittier et al. (1986) reported that if the pellets exceed a certain size (critical diameter), the lack of oxygen and nutrients within the pellet create a hollow center with a semi-anaerobic environment and low amount of viable hyphae (Figure 2.1). To avoid this, it is recommended to maintain the pellet size below a critical diameter, which is strain and cultivation condition dependent.

Table 2.1 Advantages and limitations of fungal pelleted growth in bioreactor applications.

Advantages	References
• Improvement of harvesting, good settling ability and quick separation of biomass • Improvement in culture rheology (low-medium viscosity, Newtonian flow behavior) • Lower power consumption for sufficient bulk heat and mass transfer • Biomass reuse and continuous operation of the process • Better mass exchange of oxygen and nutrients due to the decrease surface to volume ratio • Low clogging effect • Ease of scale up • High cell loading and volumetric productivities • Does not adhere to any part of the bioreactors	Metz et al., 1977; Xin et al., 2012; Rodarte-Morales et al., 2012; Sumathi et al., 2000; Renganathan et al., 2006
Limitations	**References**
• Limited transport of oxygen and nutrients within the core of the pellets, causing regions with different grow and metabolic patterns • Autolysis in the interior of large pellets • Non uniform cell aggregates	Pazouki and Panda, 2000; El-Enshasy, 2007; Calam, 1976

2.3 Fungal pelleted bioreactors for wastewater treatment

2.3.1 Potential applications and challenges

Conventionally, bacteria are the preferred microorganisms to be used in bioreactors for the treatment of wastewater, whereas the use of fungi is still in its infancy, with just few full-scale examples of fungal biotechnology. In general, bacteria outperform fungi showing better performance, not only under natural conditions, but also when applied in bioreactors. Bacteria are known to tolerate diverse conditions, grow faster and can efficiently degrade a rather broad range of pollutants as compared with fungi. On the contrary, fungi possess the ability to produce high amounts of non-specific oxidative enzymes, which allows the degradation of pollutants with highly complex structures. Moreover, fungi are able to survive under acidic conditions that allow the treatment of acidic effluents. For instance, the white-rot fungi are capable of degrading carbamazepine, one of the most studied pharmaceuticals detected in the environment and hardly removed by conventional wastewater treatment plants. Anew, fungi have also proven to be efficient in degrading carbamazepine (Jelic et al., 2012), even under non-sterile conditions (Li et al., 2015). Carbamazepine degradation can be mainly attributed to the production of fungal enzymes such as manganese peroxidase and versatile peroxidase (Golan-Rozen et al., 2011), although the cytochrome P450 also plays a key role in degradation (Golan-Rozen et al., 2011; Marco-Urrea et al., 2010).

As mentioned earlier (see section 2.2.3), fungal pellets offer a series of operational advantages when compared to the use of dispersed fungal mycelia in bioreactor applications (Table 2.1). This is mainly because of the feasibility of avoiding the operational difficulties associated with the use of dispersed mycelia growth and the easiness of separating the fungal biomass and product recovery. Moreover, better removal efficiencies have been obtained using fungal pellets compared to other forms of fungal growth for certain pollutants. Better removal efficiencies were obtained for the removal of color and aromatic compounds from pulp mill wastewater (black liquor) using *Trametes versicolor* pellets (Font et al., 2003) compared to a similar study using immobilized fungus in a fixed bed bioreactor (Font et al., 2006). Font et al. (2006) compared the laccase production and toxicity reduction correlation for both studies and observed a different behavior depending on the way *T. versicolor* was used (pellets or immobilized). The authors suggested that laccase activity was related to toxicity reduction, and that different fungal morphology might produce different laccase isoenzymes.

The suitability of bioreactors using pellets for wastewater treatment is demonstrated by its capacity to grow faster, yield high enzyme production, and associated pollutant removal. However, a series of limitations of fungal reactors have been identified, including mass transfer limitations, dead zones, preferential flow paths (by-passing), necessity to continuously remove excess biomass, high maintenance of continuous operation under stable conditions and the necessity of sterile conditions. Moreover, one of the main issues with fungal pelleted reactors consists of maintaining the fungal pellets during long-term reactor operations. After a determined period of time, dispersed growth is usually observed in the reactor, especially if no maintenance strategies are applied. This can affect the removal efficiency of the pollutant and lead to failure in bioreactor operation. These limitations can be attributed to the current lack of full-scale applications of fungal bioreactors. Hence, further research on developing full-scale systems that allow maintaining sufficient fungal growth while conserving the fungal pelleted morphology and good removal efficiency of pollutants over time needs to be carried out.

The application of fungi to treat different contaminated effluents has proven to be successful at lab-scale. Fungal pellets have been used to treat acidic olive oil washing wastewater (Cerrone et al., 2011), effluents from textile and dyestuff industries (Sanghi et al., 2006; Hai et al., 2013; Kaushik et al., 2014; Wang et al., 2015; Xin et al., 2010), stripped gas liquor effluent from a coal gasification plant, pulp, cotton and paper mill plant effluents (Taseli et al., 2004; Zhao et al., 2008), hospital and pharmaceutical effluents (Cruz-Morató et al., 2014), corn-processing wastewater (Sankaran et al., 2008) as well as effluents from agricultural activities (Mir-Tutusaus et al., 2014) and metal-containing effluents.

2.3.2 Reactor configurations

Different bioreactor configurations have been used for the growth of fungi: the stirred tank reactor being the most common one, although bubble column, airlift and fluidized bed reactor types have also been used for wastewater treatment (Figure 2.3). The most popular types of reactor configurations used with fungal pellets are described in the following subsections. The advantages and disadvantages of each reactor configuration are summarized in Table 2.2.

2.3.2.1 Stirred tank reactor

The stirred-tank reactor (STR, Figure 2.3A) is the most common reactor type for aerobic fermentations. This system consists of a tank fed with the growth medium (under sterile or non-sterile conditions) and the fungal inoculum. Air is supplied usually at the bottom of the reactor, and it is dispersed by mechanical agitation provided by an agitator and baffles (usually 4-8), assuring adequate mixing in the reactor. Agitation speed can control the fungal pellet morphology (size and shape) and therefore the enzyme activity. Under optimal mixing conditions, the STR has shown to promote and enhance the production of certain fungal enzymes. Babiĉ and Pavko (2012) observed an increased production of laccase by pellets of the white-rot fungus, *Dichomitus squalens*, when cultivated under optimal conditions in a STR with a natural inducer (saw dust). The laccase production was higher than those obtained when incubated in a bubble column reactor. This was mainly attributed to the pellet morphology obtained in the STR, with smaller, smooth and round pellets, whereas pellets in the bubble column reactor were bigger and fluffier. Similar results were obtained by Cao et al. (2014), when growing pellets of *Pycnoporus sanguineus* in a STR and in an airlift reactor. Bioreactor configuration influenced the morphology of the pellets, mainly driven by the shear forces prevailing in each reactor configuration. Pellets in the airlift reactor were fluffier than those in the STR. The hairiness nature of the pellets is highly impaired by the high shear force in the STR. The enhancement of particular enzymes by STR might then suggest the application to treat certain pollutants.

Table 2.2 Advantages and limitations of different reactor configurations using fungal pellets.

Reactor type	Advantages	Limitations
Stirred tank	• Better for viscous broths • Easy to measure and control operational parameters (pH, dissolved O_2 and shear) • Ideal for industrial applications • Size and shape of pellets can be controlled by adjusting agitation speed	• Impractical at volumes >500m^3 • Agitation power becomes too high • Shearing stress. Excessive agitation leads to the breaking off/rupture of pellets (Moreira et al., 2003)
Airlift column	• Significant reduction of energy consumption and operating costs • No focal points of energy dissipation • Shear distribution is homogeneous throughout the reactor. Low share rate • Mechanical simplicity • Suitable to processes with changing oxygen requirements (Merchuk and Gluz, 1991)	• Not suitable for viscous broths • Requirement for a minimum liquid volume for proper operation (Merchuk and Gluz, 1991) • Dead zones inside the reactor • Insufficient mixing at high biomass densities
Bubble column	• No mechanical agitation, significant reduction of energy consumption • Excellent heat and mass transfer characteristics (Kantarci et al., 2005) • Little maintenance and low operational cost • Bulk mixing and mass transfer is more cost-effective • No mechanical damage to fungal pellets (Rodríguez-Porcel et al., 2007) • Facilitates sterile operation (no moving and non-sealing parts are required) • Scale-up is relatively easy	• Short gas phase residence time • Higher pressure drop with respect to packed columns • Undefined fluid flow pattern inside the reactor • Non-uniform mixing
Fluidized bed	• Stable • Easily maintained (Andleeb et al., 2012) • Pulsed reactor • Improved version of the fluidized bed reactors • Controlled pellet growth • System allows having an active biofilm, shaving the fungal biomass when the desired thickness has been exceeded (Lema et al., 2001) • Proper fluidization by continuous supply of air, allowing gas-liquid-pellet separation (Moreira et al., 2003)	• Pellet aggregation may lead to bumping, spouting and slugging of the bed (Kunii and Levenspiel, 1968)

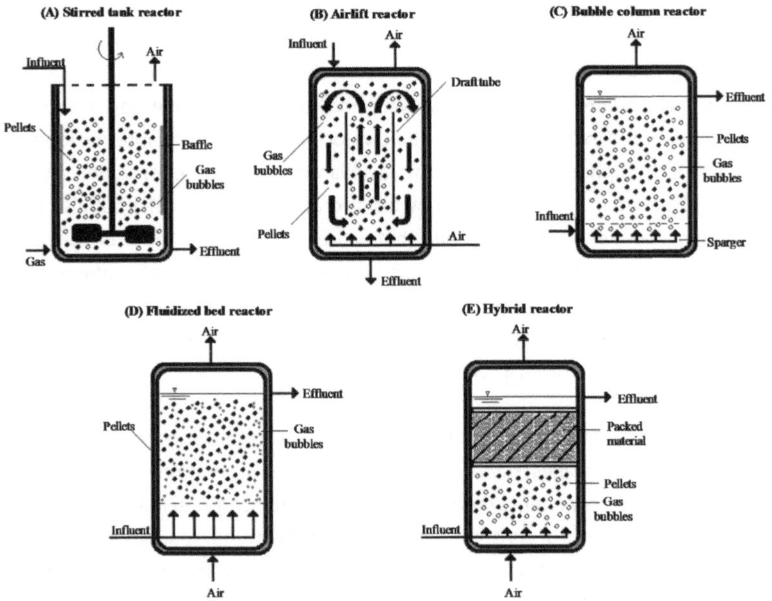

Figure 2.3 Types of reactors used for fungal pelleted growth: A) Stirred tank reactor, B) Airlift reactor, C) Bubble column reactor, D) Fluidized bed reactor, and E) Hybrid reactor.

Stirred tanks are the largest group of reactors for culturing fungal pellets and their use has also been well-documented in wastewater treatment. STR has been used for the removal of pharmaceuticals (Rodarte-Morales et al., 2012), dyes (Borchert and Libra, 2001; Sulaiman et al., 2013) and metals (Kogej and Pavko, 2001). For example, Rodarte-Morales et al. (2012) used *Phanerochaete chrysosporium* pellets in a STR for the removal of pharmaceutical compounds, *viz.*, diclofenac, ibuprofen, naproxen, carbamazepine and diazepam. High efficiencies were achieved for the removal of the anti-inflammatory agents, with only a partial removal of carbamazepine and diazepam.

Although the use of pellets in STR has been proven (Rodarte-Morales et al., 2012; Bai et al., 2003), some problems might arise: pellet size decreases with degree of agitation, with excessive agitation leading to the breaking off of hyphal fragments from the pellet surface and rupture of pellets which causes an adverse effect on enzyme secretion (Moreira et al., 2003). The next generation of STR should consider the implementation of novel mechanical systems that allow improving operational conditions in the system and therefore the removal efficiency of pollutants. As an example, Sulaiman et al. (2013) developed a STR with novel impeller geometry (60° and 180° curved blade impellers) for decolourization of triarylmethane dye by pellets of *Pycnoporus sanguineus*. High removal efficiency (>80%), at a relatively less power consumption, was achieved when using the 180° curved blade impeller along with low agitation speed compared to a traditional STR. The increase in the enzyme production and subsequently, the removal efficiency in the modified reactor were attributed to better diffusion process in the reactor.

2.3.2.2 Airlift reactor

The airlift bioreactor configuration (Figure 2.3B) was developed with an intention to prevent mechanical agitation, to ensure increased oxygen transfer rates and to provide minimal power for aeration. The bioreactor content is agitated pneumatically in a well-defined cyclic pattern of loop formation. The configuration of an airlift reactor can take either one of the two forms: (i) internal-loop or (ii) external-loop reactor. An internal airlift reactor consists of a strategically placed baffle that separates the rising medium from the downward moving medium, thereby producing adequate medium for recirculation. The average spatial liquid velocity gradients in an internal loop airlift reactor are usually constant throughout the reactor. This is advantageous for the formation and maintenance of fungal pellets (Kanai et al., 1996). An external airlift reactor comprises of a bubble column, serving as the riser section, and a downcomer section that is provided at the exterior of the bubble column. The mean density difference between the two sections enables this reactor configuration to achieve continuous recirculation and mixing of the gas and liquid phases (Merchuk and Gluz, 1991; Kanai et al., 1996). For instance, the mass transfer rate is usually affected by the mean bubble diameter, its rising velocity, disengagement and stratification.

The application of airlift bioreactors for wastewater treatment is becoming popular nowadays due to their inherent advantages such as simple design and operation, no need for mechanical agitation, low shear stress, low energy requirements for aeration, ease of maintenance, and provision to scale-up. Moreover, the costs of construction and operation of airlift reactors are considerably lower compared to other reactors, such as the STR (Fontana and da Silveira, 2012). The airlift bioreactor configuration is a suitable option for fungal pellet formation and high enzyme production. The use of pellets avoids clogging of the system due to excess biomass growth and provides better mass transfer of pollutant which is advantageous to obtain stable removal efficiencies in the bioreactor (Xin et al., 2010). Separation of the fungal biomass from the effluent is also facilitated by the use of pellets. This reactor configuration has been used for the treatment of effluents polluted with phenolic compounds (Ryan et al., 2005) and colorants (Xin et al., 2010; Rioja et al., 2008).

Even though it is difficult to compare the performance of different reactor configurations and operational conditions, the application of the airlift reactor has been shown to be suitable for the fermentation of white-rot fungi, with higher production of enzymes compared to other reactors. The laccase activity in an airlift reactor inoculated with *T. pubescens* pellets for the degradation of phenolic compounds (Xin et al., 2010) was 200 times higher than the activities observed in a bubble column reactor using *T. versicolor* pellets (Cerrone et al., 2011).

2.3.2.3 Bubble column reactor

The bubble column reactor (Figure 2.3C) belongs to the category of multi-phase reactors. It is an elongated non-mechanically agitated reactor with a gas distributor at the bottom, assuring uniform mixing and contact between the gas and the liquid phases. This type of reactor shows excellent heat and mass transfer characteristics (Kantarci et al., 2005), while allowing little maintenance and low operational costs. Bubble column reactors are quite advantageous for the use of fungal pellets, since it does not cause mechanical damage to the pellets (Rodríguez-Porcel et al., 2007). Moreover, it facilitates sterile operating conditions within the reactor by not having any moving and sealing parts. Bubble column reactors have been used for different applications with fungal pelleted forms: improved production of carotenes by *Blakeslea trispora* (Nanou et al., 2012), enhanced production of L(+)-lactic acid by *Rhizopus arrhizus* (Zhang et al., 2007), and enhanced production of lovastatin by *Aspergillus terreus* (Rodríguez-Porcel et al., 2007).

The suitability of the application of bubble column reactors for the degradation of specific contaminants present in wastewater is fungal strain dependent. Two different strains, *F. troggi* and *T. versicolor*, were grown in batch mode in a bubble column reactor to test their abilities to grow and degrade pollutants from olive washing wastewater (Cerrone et al., 2011). Growth of *F. trogii* was strongly inhibited in the reactor, compared to batch reactor operation, with nearly negligible production of laccase and rather poor removal of color (48%), COD (42%) and phenols (76%). On the other hand, *T. versicolor* showed better performance and exhibited similar behavior as when cultivated in shake flasks, with higher removal efficiencies of color (64%), COD (87%) and phenols (88%). A plausible explanation for different behavior of the fungal strains was, however, not provided. Moreover, growth in the form of pellets was conserved for the 22 d of continuous bioreactor operation.

Bubble column reactors with fungal pellets have been used to treat pollutants from different effluents, including olive washing wastewater (Cerrone et al., 2011) and cotton pulp black liquor (Zhao et al., 2008). The use of bubble column reactors enhances the production of selective fungal enzymes compared to other reactor configurations. Babiĉ and Pavko (2012) observed enhanced production of manganese peroxidase by pellets of *D. squalens* using a bubble column reactor when compared to using a STR. This was mainly attributed to the pellet morphology obtained in the bubble reactor (big and fluffy).

2.3.2.4 Fluidized bed reactor

The fluidized bed reactor (Figure 2.3D) is characterized by its plug flow nature of fluid movement inside the reactor. The fluidization occurs when solid material (*i.e.*, biomass) is suspended in an upward-flowing stream of fluid, which can be either liquid or gas. The fluid velocity in the reactor is good enough to maintain the particles in suspension without getting washed out from the system. The solid material swirls around the bed rapidly, creating uniform dispersion inside the reactor, which offers a large surface area for higher mass transfer and better mixing. The use of fungal pellets might cause poor fluidization due to pellet aggregation, leading to spouting or slugging beds (Glicksman, 1999). Strategies like partial biomass purging allow avoiding such operational issues, ensures proper fluidization of the bed and offers good stability to variations in pollutant loading rates and HRTs (Rodríguez-Rodríguez et al., 2012). An improved version of the fluidized-bed is the pulsing bioreactor, which avoids the drawbacks of conventional fluidized bed reactors by controlling the pellet growth with a system that allows not only to maintain active biomass, but also to "shave" the fungal biomass when the desired thickness has been exceeded (Lema et al., 2001). The formation of compact pellets as spheres is promoted by the air pulses (Rodarte-Morales et al., 2012), along with their homogeneous distribution in the reactor (Cruz-Morató et al., 2013). The fluidized bed reactor has been used for the removal of pharmaceutical active compounds (Jelic et al., 2012; Cruz-Morató et al., 2014), endocrine disrupting contaminants (Blánquez et al., 2008), dyes (Baccar et al., 2011), colorants, COD, among others (Font et al., 2003; Jaganathan et al., 2009).

2.3.2.5 Hybrid reactors

Hybrid reactors combine the strengths of both pelleted and attached growth in a single reactor (Figure 2.3E). Hybrid reactor designs enable the possibility to benefit from the use of reactors that on their own might not be suitable for fungal pelleted growth. The fungal pellets can be present in the bottom reactor compartment and a fixed-bed section (bacterial or fungal) in the upper reactor compartment. Inclusion of the fixed-bed section provides an effluent polishing step or support onto which non-pellet forming fungal species can colonize the bioreactor. Moreover, such a fixed biomass section can be more resistant to substrate toxicity and offer high resistance to shock loads (Rodriguez and Couto, 2009).

Hai et al. (2013) used a stirred tank membrane bioreactor with pelleted and attached growth of *T. versicolor* for the removal of the azo dye acid orange 7 under non-sterile conditions. The synergetic effect of both types of fungal growth provided a more stable and longer reactor performance with high enzymatic activity and decolourization compared to a similar reactor with stand-alone fungal pellets. The authors assessed the importance of favoring fungal biofilm growth in the reactor by purposely damaging it. After disturbing the biofilm, the performance of the reactor gradually decreased and it was prone to bacterial contamination. A control reactor containing only attached growth of the biocatalyst was not performed in this study. Besides, complete removal of organic matter was also not achieved, a common issue that has been reported during the long-term operation of fungal bioreactors. Therefore, the requirement of a post-treatment step was suggested by the authors in order to ensure complete degradation of organic matter.

2.3.3 Reactor design for fungal pelleted reactors

The main design criteria for reactors with fungal pellets should consider adequate transfer of oxygen and other nutrients, low shear stress and adequate mixing (Wang and Zhong, 2007; Zhou et al., 2010). Additionally, design conditions should also ensure adequate nutrient supply to the cells and prevent waste products/metabolic end-products from accumulating inside the pellets. For an effective scale-up of any optimized bioreactor with fungal pellet biomass, transport phenomena should be studied together with cell growth and pollutant degradation kinetics. Operating parameters, such as temperature, pH, dissolved oxygen, hydraulic retention time and substrate concentration should be optimum to maintain a proper structure and function of the pellets.

2.3.3.1 Mixing

Mixing is a physical operation which reduces non-uniformity in a fluid by eliminating concentration gradients of the nutrients, gases, temperature or pH (Bailey and Ollis, 1986). Mixing is also crucial for efficient heat and mass transfer; it depends on a series of parameters, including power input, vessel shape and size, fluid hydrodynamics, characteristics of the agitation mechanism (rotation speed, air flow rate and geometry of the impeller), rheology of the broth, growth rate and biomass concentration (Cascaval et al., 2004; Núñez-Ramirez et al., 2012). Particularly morphology plays a very important role on the mixing efficiency in fungal reactors. Viscosity of the fungal broth is determined by the morphology of the fungus, which dictates the rheological behavior of the reactor. Cascaval et al. (2004) observed that the influence of the rotation speed is higher for the pelleted form due to the apparent low viscosity compared to dispersed mycelia.

Different regimes can be found during the mixing process: laminar, transitional or turbulent, which are mainly restricted by the broth viscosity in the reactor. In general, the laminar and the transitional regimes are the preferred ones for biological systems (Núñez-Ramirez et al., 2012). A small mixing time is highly desired for a better and faster distribution of oxygen, substrate and temperature to the cells inside the pellets for enhanced growth and performance. In bioprocesses that are typically sensitive to shear-force fluxes, mixing could be improved by altering the aeration rates or agitation speed in the bioreactor. Nevertheless, mixing has to balance with the increase of shear stress because shear force caused by intense aeration or agitation can damage cell pellets or cause pellet erosion.

2.3.3.2 Oxygen transfer

Oxygen transfer often becomes a limiting step to the optimal performance of biological systems and also for scale-up because oxygen is only sparingly soluble in aqueous solutions. When the supply of oxygen is limited, both pelleted growth and bioreactor performance can be severely affected (Jesús et al., 2009). It has been reported that the microbial oxygen requirement is related to the synthesis of enzymes responsible for the breakdown and utilization of the substrate, cell maintenance energy, as well as electron acceptor required for growth and metabolism of the microorganism (Jesús et al., 2009). Oxygen conditions influence the morphology and size of the fungal pellets (Bermek et al., 2004) and therefore the enzyme activity and pollutant removal capacity. Under oxygen-rich conditions, more homogeneous pellets in both size and shape appear, as slightly smooth spheres, leading to an earlier and enhanced production of certain enzymes (Bermek et al., 2004).

2.3.3.3 Shear force

Fungal cells are more sensitive to shear stress and shear-induced cell death has been reported in fungal strains (Zhong et al., 2010). The prevalence of various shear forces in reactors also acts as a controlling parameter of fungal morphology; however, these forces can also cause damage to the cells, especially at high energy dissipation rates. The development of fungal branches (hairiness) of pellets is highly impacted and it can be impaired by the shear force, which depends on the reactor type and geometry. According to literature reports, different sizes and shapes of fungal pellets have shown to exhibit different enzyme production rates in STR, bubble column and airlift reactor (Babiĉ and Pavko, 2012; Cao et al., 2014) and this was attributed to the different shear forces in these different bioreactor configurations. Sparging can sometimes cause shear damage, which can occur at different locations in the fluidized bed reactor: at the bubble generation zone, the rising zone and the bubble disengagement zone (Jordan et al., 1994; Papoutsakis, 1991). Aeration at a single port at a high rate creates high shear stress that hampers the self-immobilization of biomass. However, a fragmented approach of aeration, *i.e.*, separate aeration zones can reduce shear stress (Fassnacht and Portner, 1999). This can be achieved by a cross flow design where aeration is provided at multiple points along the length of a fixed-bed reactor, which as well reduces the pressure drop enabling the use of the flow rate and even flow distribution across the bed (Ajmera et al., 2002). To overcome the problem of the rising air (O_2) bubbles, diffusion controlled (bubble-less) aeration using membranes can be employed (Kretzmer, 2002).

2.3.4 Sterile versus non-sterile conditions

The application of environmental biotechnologies has been centered traditionally on the use of non-sterile systems. Most of the studies in the literature about the use of fungal reactors report experimental data performed under sterile conditions. However, it raises concerns about the treatment of real wastewater under non-sterile conditions. The robust design of bioreactors to maintain similar performance under sterile and non-sterile conditions is often a challenging task that has not been addressed adequately in the literature (Rene et al., 2010). Non-sterile conditions are simulated in lab-scale studies by using non-sterilized tap water instead of distilled water, the use of industrial grade chemicals instead of reagent grades, and by allowing direct air contact of the reactor content which is usually achieved by leaving the reactor open. The development of symbiotically striving fungal-bacterial consortia which are able to remove toxic pollutants from wastewater, at the same time surviving and outlasting the colonization of indigenous bacteria in wastewater, is a promising technology.

One of the main concerns of working under non-sterile conditions is the possibility of bacterial contamination in the reactor. Bacterial colonization will occur if no special control of bacterial intrusion (*i.e.*, sterilization) is performed, particularly during the start-up period. The presence of bacteria in the fungal system may create competition for the substrate, provoke disruption of the fungal growth, damage the fungal mycelium (Rene et al., 2010) or reduce the expression of the fungal enzymes (Libra et al., 2003; Gao et al., 2008; Yang et al., 2013), ultimately even leading to the deterioration of the fungal activity. For instance, Hai et al. (2013) reported the gradual decline in enzymatic activity caused by fragmentation of *Coriolus versicolor* (also known as *Trametes versicolor*) fungal pellets when grown in a stirred tank reactor for the removal of dyes after four weeks of operation and bacterial colonization. Similar observations have been reported during the treatment of wastewater polluted with dyes and anti-inflammatory compounds (Yang et al., 2013; Blánquez et al., 2008). Therefore, there is an urge to develop technologies that allow increasing the fungal competitiveness over bacteria and that can suppress bacterial colonization.

Different strategies that have been suggested in the literature to suppress bacterial growth under non-sterile conditions are:

i) *Immobilization of fungal cultures*: Immobilization of fungi onto carriers such as polyurethane foam restrain the growth of diverse bacteria (*e.g.*, *Coccus* and *Bacillus* spp.) under non-sterile conditions, allowing to produce higher amounts of fungal enzymes (Gao et al., 2008). In another study, under non-sterile conditions, bacterial colonization was prevented from interfering with the performance of *C. versicolour* (NBRC 9791) for a short period of time (~26 days) in a sequencing batch reactor treating wastewater containing acid Orange 7 (Hai et al., 2013). However, after 26 days, bacterial contamination was observed, that increased the viscosity of the medium and reduced the color removal efficiency.

ii) *Reduction of medium pH*: Fungi grow and reproduce optimally under acidic conditions. In contrast, bacteria preferentially grow in a pH near neutrality. Abundance of fungal species over bacteria in non-sterile reactors is mainly attributed to the acidic conditions (Libra et al., 2003). A decrease of the medium pH may temporarily suppress bacterial growth. However, this strategy is not a long-term protection against bacterial contamination, since bacterial populations may be capable of adapting to acid conditions (Libra et al., 2003).

iii) *Limitation of nitrogen in the medium*: Nitrogen-limited media not only favor fungal growth in pelletized morphology, but also suppress bacterial growth. Usual restrain of bacterial growth due to nitrogen limitation occurs during the start-up of a bioreactor. However, with operational time, bacterial colonization may occur on the fungal mycelium which in turn acts as the carbon and nitrogen source for bacteria (Libra et al., 2003).

iv) *Selective disinfection*: This refers to the elimination of undesired microorganisms in the system without suppressing fungal cells. The type of disinfectant and dose required depend on a specific type of undesired microorganism and the wastewater characteristics. The use of bactericides such as ozone has proven to limit the growth of contaminating microorganisms in fungal reactor systems (Cheng et al., 2013; Sankaran et al., 2008). The use of ozone not only suppresses bacterial growth, but also offers better reactor stability and facilitates the removal of pollutants in wastewater (example: Acid Blue 45) (Cheng et al., 2013). Sankaran et al. (2008) observed inhibition of bacterial population during fungal cultivation on corn-processing wastewater. However, the replacement of *Rhizopus oligosporus*, the initial species used for inoculation, by *Thricosporon beigelii*, a wild type of filamentous yeast, was also observed.

v) *Microscreen*: This selection system allows the separation of bacteria from the fungal biomass. The fungal biomass is retained in the system, while the bacteria are washed out with the effluent (van Leeuwen et al., 2003). By controlling the retention time of the fungal biomass (SRT), fungal dominance can be achieved (van der Westhuizen et al., 1998).

2.3.5 Biomass recycle in fungal pelleted reactors

For environmental engineering applications, recycling of fungal pellets could be both biologically challenging and technically demanding considering the amount of residual dye and heavy metals that are concentrated when recycling the pellets. Although there are reports that have focused on using fungal pellets to treat industrial wastewater containing dyes, heavy metals, organic and inorganic pollutants, the recycle of used pellets is hardly exploited under the tested conditions. Table 2.3 gives a summary of the reuse ability of some fungal pellets for dye wastewater treatment. The longevity of fungi-inoculated bioreactors that operate with pellet recycle depends on factors such as enzyme activity, surface charge, frequency of spore formation, the presence of hydrophobic proteins, ability to re-acclimatize, and the pollutant load applied to the bioreactor.

Concerning the recycling of used fungal pellets, gravity settling as in the activated sludge process is not sufficient and it generally requires one or more solid-liquid separation steps. As shown schematically in Figure 2.4, the maintenance phase for recovery and reuse consists of a settle, draining, re-suspension, draining and filling step, wherein each phase requires at least a few hours in order to maintain the quality of the freshly rejuvenated pellets to receive the next inflow in continuous mode. After settling the fungal pellets, the treated effluent is withdrawn and the settled fungal pellets are washed with tap water to remove the dispersed and dead biomass and other water-soluble end-products. After draining the wash medium, the new batch of wastewater is fed to the bioreactor. During this step, process parameters (pH, temperature, oxygen flux) favoring fungal growth are automatically provided.

High biomass activities can be maintained by controlling the injection of required micro and macro nutrients, by inoculating the bioreactor with new mycelia periodically and by maintaining the speed of mixing through aeration. This biologically engineered operational mode is similar to the operational cycle of a conventional sequencing batch reactor consisting of a fill, react, settle and decant phase: each of these steps being controlled in well-defined time sequences.

Table 2.3 Reuse ability of some fungal pellets for dye wastewater treatment.

Fungal species	Dye stuff	Concentration (mg L^{-1})	Decolourization (%)	Reuse cycles	References

Aspergillus lentulus	Acid Blue 120	100	90	5	Kaushik et al., 2014
Trametes versicolour	Blue 5 Red 198	100	97-99.5	18	Borchert and Libra, 2001
	Remazol Brilliant Blue R,	500	91-99		
	Acid Violet 7	100-500	>95	9	Zhang and Yu, 2000
	Black Dycem	150	86-89		Baccar et al., 2011
	Remazol Brilliant Violet	100	80	8	Yesilada et al., 2002
Basidiomycete	Orange II	1000-2000	>95	9-27	Bayramoğlu et al., 2009
Funalia trogii	Astrazon Red	13-264	28-95	5	Harms et al., 2011
Penicillium janthinellum	Congo Red	150	>99	5	Fu et al., 2001

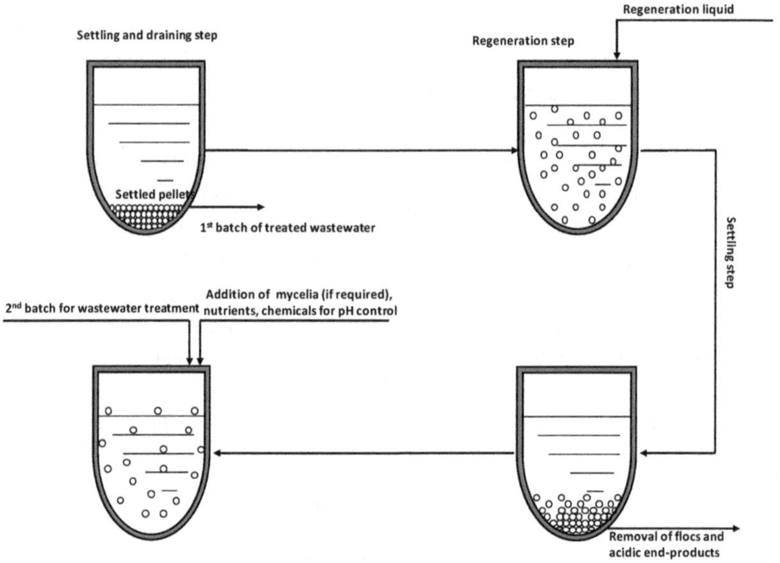

Figure 2.4 Illustration of biologically engineered maintenance phase for the reuse of fungal pellets treating wastewater.

2.4 Removal of organic and inorganic pollutants

2.4.1 Removal of organic pollutants

The unique capability of fungi to produce large amounts of specific and non-specific enzymes makes fungi very attractive for the degradation of complex organic pollutants. Diverse fungal strains have been used to remove three major groups of organic pollutants: (i) synthetic dyestuffs - widely used in textile, printing and paper industries, usually containing high amounts of COD, BOD, suspended solids, toxic compounds and color, (ii) phenols - mainly present in wastewaters from pharmaceuticals, pesticides, solvents, paper and pulp industries, and (iii) pharmaceutical compounds (*i.e.*, anti-inflammatory agents, analgesics, antibiotics, psychiatric and anticancer drugs) - these active organic micro-pollutants are of great concern because of the inefficiency of traditional secondary wastewater treatment plants to remove these pollutants.

Different fungal mechanisms to remove organic pollutants from wastewater have been proposed. The most common ones are enzymatic degradation, bioaccumulation and biosorption. The major difference compared to the removal of heavy metals is the main role of the metabolically active uptake of organic compounds in the removal of such pollutants (Murugesan and Kalaichelvan, 2003). The ability of certain fungal strains to oxidize a wide variety of organic compounds is attributed to the high production of oxidative enzymes such as lignin peroxidase, manganese peroxidase and laccase (Cabaleiro et al., 2002). A detailed list of major classes of enzymes involved in the fungal catabolism of organic pollutants has been reported by Harms et al. (2011).

2.4.1.1 Removal of dyes

Concerning dye removal from wastewater, decolorization using fungi is mainly attributed to the presence of laccase, manganese peroxidase and ligning peroxidase, although biosorption can also play an important role in the decolourization process (Baccat et al., 2011; Harms et al., 2011). Fungal decolourization of dye wastewaters has been extensively reviewed (Kaushik et al., 2014; Zhang et al., 2011), particularly, the use of white-rot fungi (Rodarte-Morales et al., 2012). Fungal pellets possess an adaptive feature that allows them to survive and efficiently decolourize high dye concentrations after consecutive use. Moreover, fungal biomass can also serve as a good biosorbent for the removal of organic compounds (Baccar et al., 2011). The removal of organic pollutants with fungi can be performed with live, dead or inactivated biomass, which also dictates the specific removal mechanisms. Xin et al. (2012) demonstrated that different mechanisms were involved in the removal of Acid Brilliant Red B, depending on the use of dead or alive pellets of *Trichoderma* sp. It was observed that the main mechanism for the living pellets was bioaccumulation, principally in the cell wall and fungal cytoplasm. On the other hand, monolayer adsorption onto the surface of the pellets was the removal mechanism when dead pellets were used.

Table 2.4 summarizes important information from recent literature reports (last decade), wherein different process conditions have been used to test the removal efficiency of organic pollutants from water using fungal pelleted reactors. Most of the research has been done for the degradation of dyes, although only some investigations have particularly focused on the development of fungal bioreactors that could be scaled-up and used for this purpose, for which the airlift (column) and stirred tank reactor types are the most popular ones. Particular success was achieved in an airlift column reactor using pellets of *P. oxalicum*. In this fungal pelleted reactor, complete removal of Cu-complex Reactive Blue 21 dye (200 mg L^{-1}) was achieved in ~12 h. The results from that study showed rapid bioaccumulation of the dye complex and the fungal pellets were largely unaffected when exposed to high dye concentrations (Xin et al., 2010).

Long-term operation of fungal pelleted reactors to remove dyes has been successfully achieved under sterile conditions. A STR using pellets of *T. versicolor* was able to remove (91-99%) up to 500 mg

L^{-1} of azo dyes for as long as 28 weeks (Borchert and Libra, 2001). However, as mentioned earlier, long-term operation under non-sterile conditions is still a challenge. In that study, under non-sterile conditions, the system was stable only for 5 cycles (Borchert and Libra, 2001). A bubble column reactor with pellets of *P. ostreatus* was also proved as a suitable option for the decolourization of cotton pulp black liquor (Zhao et al., 2008). The reutilization of the fungal pellets allowed the maximal removal efficiency of 76% of color from the effluent. The system was run in repeated batch modes, for 4 cycles of 12 days each. The COD removal efficiency was around 70%. High COD values were attributed to high content of lignin and other organic compounds in the black liquor. Besides, 10 g L^{-1} of glucose was added to the influent in order to promote fungal growth. The presence of high amounts of residual COD in the effluent is a common issue during fungal treatments. In such cases, post-treatment of the effluent is required to reach desired removals of COD.

2.4.1.2 Removal of phenolic compounds

Some fungal strains, particularly white-rot fungi, are capable of degrading monomeric and polymeric phenolic compounds, driven by oxidative enzymes, including lacasses, tyrosinases, and manganese peroxidases (Cerrone et al., 2011; Zhang et al., 2010). Biosorption has also been observed as a removal mechanism for these compounds (Filipović-Kovačević et al., 2010; Bayramoğlu and Arica, 2008). More than 40 white-rot fungal strains, some of them being pellet-forming organisms, were identified for their ability to reduce phenol contents (>60%) from olive mill wastewater, mainly correlated to the secretion of lacasses and peroxidases (Zhang et al., 2010).

Cerrone et al. (2011) demonstrated the ability of *T. versicolor* pellets to degrade pollutants from diluted olive washing wastewater in a bubble column reactor. Maximum removal efficiencies (color 60%, COD 72%, phenols 87%) were obtained after 8 d in batch operating mode and the reactor was thereafter operated for 22 d in continuous mode. The removal of nutrients was not completed achieved and the production of laccase was considerably low (52 U L^{-1}). The use of the treated effluent for soil ferti-irrigation was recommended by the authors. Under the optimized conditions, this system (10 m^3 working volume) would allow to treat the effluent produced by the washing machines used in olive mills on a daily basis (Cerrone et al., 2011). Ryan et al. (2005) studied the removal of phenols from stripped gas liquor effluent with *T. pubescens* pellets in an airlift reactor. The system was operated for 15 d demonstrating high phenol removal rates (0.033 g L^{-1} d^{-1}) and high production of laccase (11.8 U mL^{-1}). Such high phenol removal rates suggest the potential of using this system for the removal of phenolic compounds.

Table 2.4 Fungal pelleted reactors for the removal of organic pollutants.

Fungal strain	Reactor type	Pollutant	Operational conditions	Pellet description	Pollutant concentration and removal efficiency	References
Aspergillus lentulus	Pneumatic aerated bioreactor	Dye • Acid Blue 120	Fed-batch Inoculum: 2 g dry biomass L^{-1} Working volume: 3.5 L pH 6.5, T: 30 °C Operation time: 5 cycles (12 h each)	Not reported	Concentration 100 mg L^{-1} Removal 90%	Kaushik et al., 2014
Aspergillus niger	Column	Aniline COD	Sequencing batch Inoculum: 2364 mg L^{-1} MLSS Aeration: air 2 L h^{-1} pH: 7-7.2, T: 30°C HRT: 20 h Electron donor: glucose	Fungal pellets were used as biomass carrier for immobilization of aniline-degrading bacterium	Concentration 98-142 mg L^{-1} Removal 0.9 mg aniline L^{-1} d^{-1}	Zhang et al., 2011
Phanerochaete chrysosporium	Stirred tank	Pharmaceuticals • Diclofenac (D) • Ibuprofen (I) • Naproxen (N)	Fed-Batch Inoculum: 1.2 g L^{-1} Working volume: 2 L Agitation: 200 rpm Aeration: air/periodic O_2 pulse pH: 4.5, T: 30°C HRT: 24 h Electron donor: glucose Operation time: 30 d	Oxygenated system: 2.8 mm Ø. Dense, hyphae-free, compact spheres Aerated system: 2.8 mm Ø. After 7 d, excessive growth, hyphal branches and aggregates	Concentration 0.6-1.2 mg L^{-1} Removal D: 100% N: 77-99% I: 100%	Rodarte-Morales et al., 2012
	Stirred tank	Pharmaceuticals • Diclofenac (D) • Ibuprofen (I) • Naproxen (N) • Carbamazepine (C)	Working volume: 1.5 L Aeration: air, 0.5-3 L min^{-1} pH:4.5 T: 30 °C HRT: 24 h Electron donor: glucose Operation time: 50 d	Formation of hyphal branches (9.3 mm) after 10 d. Smaller pellets (7 mm) were formed (30-50 d)	Concentration D, I, N: 1 mg L^{-1} C: 0.5 mg L^{-1} Removal D: 92% I: >95% N: >95% C: 24-63%	Rodarte-Morales et al., 2012

Fungal strain	Reactor type	Pollutant	Operational conditions	Pellet description	Pollutant concentration and removal efficiency	References
Phanerochaete chrysosporium	Open tank	Dye • Acid Blue 45	Continuous, non-sterile, O_3 treatment unit, 3 reactors (R1, R2 and R3) Working volume: 2.7 L Inoculum: R1: 1.7 g L^{-1} free mycelium as pellets R2: 1.7 g L^{-1} immobilized mycelium with knotted cotton thread carriers R3: 4.3 g L^{-1} same as R2 Aeration: air 2.7 L min^{-1} pH: 4.5, T: 25°C Electron donor: glucose HRT: 3 d Operation time: R1: 12 d; R2, R3: 25 d	R1: initial size 3-5 mm Ø After 12 d, high number of hyphal fragments R2: immobilized carriers, thickness 1-8 mm R3: immobilized carriers, thickness 1.5-2 mm	Removal R1: 55% R2: 64% R3: 84% O_3 unit: 64.8%	Cheng et al., 2013
Penicillium oxalicum	Airlift	Dye • Cu-complex Reactive Blue 21	Batch, non-sterile. Inoculum: 2% (v/v) Aeration: air 0.6 L min^{-1} pH: 5.5, T: 30 °C Electron donor: glucose Operation time: 12, 24, 48 h	12 h - 1.52 mm Ø 24 h - 2.48 mm Ø 48 h - 3.83 mm Ø	Concentration 200 mg L^{-1} Removal 100%	Xin et al., 2010
Pleurotus ostreatus	Bubble column	Pollutants from cotton pulp black liquor • Colorant • COD	Sequencing batch Inoculum: 5.0 g dry weight Aeration: air, 1.2 L min^{-1} pH 6.0; T: 30°C Electron donor: glucose Operation time: 48 d	Not reported	Concentration Color: 19,000-20,000 CU COD: 4000-5000 mg L^{-1} Removal Colour: 76% COD: 80%	Zhao et al., 2008
Trametes (Coriolus versicolour)	Bubble column	Pollutants from olive washing wastewater • Colorant • COD • Phenols	- Batch Inoculum: 100 mL mycelium Working volume: 1 L Aeration: air 0.3 vvm pH: 4.1-4.3, T: 28°C Operation time: 8 d - Continuous Same as batch Operation time: 22 d	5-7 mm Ø	Concentration Colorimetric units: 385 COD: 2500 mg L^{-1} Phenols: 277 mg L^{-1} Removal Batch, continuous Colour: 60%, 65% COD: 72%, 73% Phenols: 8%, 89%	Cerrone et al., 2011

Fungal strain	Reactor type	Pollutant	Operational conditions	Pellet description	Pollutant concentration and removal efficiency	References
Trametes (Coriolus versicolour)	Stirred tank-membrane	Dyes • Azo dye Acid Orange 7	Continuous sequencing batch Inoculum: 0.5 g Working volume: 1.5 L Agitation: 35 rpm Electron donor: starch Non-sterilized Reactor with membrane: Attached growth + fungal pellets	0.7 cm Ø, conversion to large irregular clumps	Concentration 100 mg L^{-1} Removal ~ 95%	Hai et al., 2013
	Stirred tank	Dyes • Reactive Black 5 • Reactive Red 198 • Reactive Blue 19	Sequencing batch Inoculum: 0.5 L Working volume: 4 L Agitation: 200 rpm Aeration: oxygen pH: 5, T: 18-24°C Electron donor: glucose Operation time: 28 weeks	1-2 mm Ø. Small amount of mycelial mats were also found	Concentration 100 or 500 mg L^{-1} Removal 91-99%	Borchert and Libra, 2001
	Up-flow air pulsed	Dye • Grey LanasetG	Batch and continuous Synthetic and real industrial wastewater Inoculum : 3.2g L^{-1} Working volume: 10 L Aeration: air pulses (0.16 s^{-1}) pH: 4.5, T: 25°C HRT: 48 h Electron donor: glucose Operation time: 3 months - sterile solution >70 d - non-sterile solution 15 d - real wastewater (non-sterile)	Not reported	Concentration 150 mg L^{-1} Removal Synthetic solution (no disturbance) > 90% Synthetic solution (adapted to industrial disturbances) 78% Real industrial wastewater (non-sterile) 40-60%	Blánquez et al., 2008
	Fluidized bed	Endocrine disrupting contaminants • 17β-estradiol (E1) • 17α-ethynylestradiol (E2)	Continuous Aeration: air 3L h^{-1} Inoculum: 3.2 g L^{-1} HRT: 120 h pH: 4.5, T: 22 ± 2°C Electron donor: glucose Operation time: 26 d	Not reported	Concentration E1: 18.8 mg L^{-1} E2: 7.3 mg L^{-1} Removal 97-99% E1: 0.16 mg L^{-1} h^{-1} E2: 0.09 mg L^{-1} h^{-1}	Blánquez and Guieysse, 2008

Fungal strain	Reactor type	Pollutant	Operational conditions	Pellet description	Pollutant concentration and removal efficiency	References
Trametes (Coriolus versicolour)	Fluidized bed	Tannery dye (Black Dycem)	Batch and sequential batch Working volume: 1.5 L Aeration: air 0.16 s^{-1} Inoculum: 5 mL of mycelial suspension pH: 4.5, T: 35°C Electron donor: glucose Operation time: Batch - 96 h Sequential batch - 3 batches of 36 h each	Not reported	Concentration: 150 mg L^{-1} Removal: Batch: 97% Sequencing batch: 86-89%	Baccar et al., 2011
	Fluidized bed	Carbamazepine	- Batch Inoculum: 3.8 g dry wet pellets Working volume: 1.5 L Aeration: air pulses 12 L h^{-1} pH: 4.5, T:25°C Electron donor: glucose - Continuous Same as batch HRT: 3 d	Not reported	Concentration: 0.2 mg L^{-1} Removal: - Batch: 96% after 2 d - Continuous: 54% after 25 d	Jelic et al., 2012
	Fluidized bed	Pharmaceuticals • Analgesics • Anti-inflammatories • Antibiotics • β blockers • Psychiatric drugs	Batch Inoculum: 1.5 g L^{-1} Working volume: 10 L Aeration: air pulses pH 4.5, T: 25°C Electron donor: glucose Real wastewater Operation time: 8 d	Not reported	Concentration: In the range of μg L^{-1} Removal: 100%	Cruz-Morató et al., 2013
	Fluidized bed	Pharmaceuticals	Batch Inoculum: 1.4 g dry biomass L^{-1} Aeration: air pulses pH 4.5; T: 25°C Electron donor: glucose Hospital wastewater, sterile and non-sterile conditions	Under non-sterile conditions, after 5 d, breakdown of the fungal pellets into free mycelium was observed	Total amount: -Sterile: 8185 μg -Non-sterile: 8426 μg Removal: -Sterile: 83.2% -Non-sterile: 53.3%	Cruz-Morató et al., 2014

Fungal strain	Reactor type	Pollutant	Operational conditions	Pellet description	Pollutant concentration and removal efficiency	References
Trametes (Coriolus versicolour)	Fluidized bed	Pharmaceuticals • X-ray contrast agent iopromide (IOP) • Fluoroquinolone antibiotic ofloxacin (OFLOX)	Batch Inoculum: 1.4 g dry biomass L^{-1} Aeration: air pulses, 12 L h^{-1} pH 4.5; T: 25°C Electron donor: glucose Hospital wastewater, sterile and non-sterile conditions Operating time: 8 d	Not reported	Concentration: -Sterile: IOP 105 μg L^{-1}, OFLOX 32 μg L^{-1} -Non-sterile: IOP 419.7 μg L^{-1}, OFLOX 3.3 μg L^{-1} Removal: -Sterile: IOP 87%, OFLOX 98.5% -Non-sterile, IOP 65.4%, OFLOX 99%	Gros et al., 2014
	Fluidized bed	Pharmaceutical • Clofibric acid	Continuous Inoculum: 3.8 g dry biomass Aeration: air pulses, 12 L h^{-1} pH 4.5; T: 25°C; HRT: 4 d Working volume: 1.5 L Electron donor: glucose Operating time: 24 d	Not reported	Concentration: 169 μg L^{-1} Removal: 80%	Cruz-Morató et al., 2013
	Fluidized bed	Anticancer drugs • Ciprofloxacin (C) • Tamoxifen (T)	Batch Inoculum: 2.0 g dry biomass L^{-1} Aeration: air pulses, 12 L h^{-1} pH 4.5; T: 25°C Electron donor: glucose Hospital wastewater, sterile and non-sterile conditions	Not reported	Concentration: C:7000 ng L^{-1} T: 970 ng L^{-1} Removal: C: 84% T: 91%	Ferrando et al., 2015
	Airlift	Colorants	Continuous Inoculum: 1.2 g L^{-1} Working volume: 7.6 L Aeration: air 1.6vvm pH 4.5, T: 30 °C Electron donor: glucose Real wastewater Operation time: 23 d	Pellet morphology was maintained during operation. Pellets turned brown, probably after exposition to the colorants	Concentration: 2.0 au at 475 nm absorbance Removal: 60%	Rioja et al., 2008

Fungal strain	Reactor type	Pollutant	Operational conditions	Pellet description	Pollutant concentration and removal efficiency	References
Trametes pubescens	Airlift	Phenolic compounds	Batch Inoculum: 350 mL of mycelial suspension Working volume: 3.5 L pH 5.0 Aeration: air 2.2 L min^{-1} Operation time: 15 d Electron donor: glucose	Shape and size changed with phenol concentrations. At high concentrations, pellets were irregularly shaped, smaller and less hairy	Concentration: 3.45 mM Removal: 0.033 g L^{-1} d^{-1}	Ryan et al., 2005

Note: COD- Chemical oxygen demand; HRT- Hydraulic retention time; MLSS- Mixed liquor suspended solids.

2.4.1.3 Removal of pharmaceutical compounds

In the last decades, there has been an increase in the number of fungal reactor studies for the removal of pharmaceuticals from water. Particularly, systems with fungal pellets have been used to remove pharmaceuticals including analgesics, anti-inflammatories, β-blockers, psychiatric drugs, antibiotics and anticancer drugs (Jelic et al., 2012; Marco-Urrea et al., 2010; Cruz-Morató et al., 2014; Rodríguez-Rodríguez et al., 2012; Cruz-Morató et al., 2013; Farkas et al., 2013; Bosso et al., 2015), as well as other toxic compounds coming from hospital wastewater such as radiocontrast agents (Farkas et al., 2013). Removal of pharmaceuticals is associated to hydroxylation, formylation, deamination and dehalogenation processes catalyzed by fungi (Cruz-Morató et al., 2013). The removal mechanisms and possible degradation pathways for trace organic contaminants have already been discussed (Yang et al., 2013). Removal of diverse pharmaceuticals using fungal pelleted reactors has been successfully achieved under non-sterile conditions from urban (Cruz-Morató et al., 2013) and hospital wastewaters (Farkas et al., 2013; Bosso et al., 2015). A batch fluidized bed reactor containing *T. versicolor* pellets was used for the removal of pharmaceuticals from non-sterile urban wastewater at their environmental concentrations (Cruz-Morató et al., 2013). Seven out of ten compounds detected were totally removed. A similar study was reported for the removal of pharmaceuticals from hospital wastewater (Cruz-Morató et al., 2014). From the 51 initially detected pharmaceutically active compounds, 46 were removed, either partially or completely, from the wastewater.

It should be highlighted that compounds that are normally not efficiently removed by conventional wastewater treatment plants, such as diclofenac and carbamazepine, are successfully removed in fungal pelleted reactors (Cruz-Morató et al., 2014; Jelic et al., 2012). Carbamazepine, an active compound to treat epileptic seizures, nerve pain and mental illnesses, is barely removed (~10%) in wastewater treatment plants (Pakshirajan et al., 2009). This pharmaceutical was successfully removed (96%) in batch conditions using *T. versicolor* pellets in a fluidized bed reactor (200 $\mu g\ L^{-1}$ initial concentration) after 2 d (Jelic et al., 2012). However, under continuous operation, the system was not efficient (54%), even after 25 d of operation (Jelic et al., 2012). The authors did not provide an explanation for the observed decrease in removal efficiency.

Long-term, stable performance of fungal pelleted bioreactors is always desired. Rodarte-Morales et al. (2012) compared the performance of a stirred tank reactor and a fixed-bed reactor using *P. chrysosporium* pellets for the removal of diclofenac, ibuprofen, naproxen, carbamazepine and diazepam. Both reactors showed good stability for over 50 d, with high removal efficiencies. Almost complete removal of the anti-inflamatories was achieved in both reactors; however, the removal of carbamazepine and diazepam was better in the fixed-bed reactor when a continuous air flow was used.

The treatment of real hospital wastewaters under non-sterile conditions with fungal pelleted reactors is promising. In fact, for some specific pollutants, the removal efficiency was higher under non-sterile conditions than those observed using controlled sterile effluents (Farkas et al., 2013; Bosso et al., 2015). This could be explained by a possible synergetic effect of fungi and bacteria present in wastewater for the degradation of the pollutants. Further investigation on this domain of research should address the use of multi-species bioreactors in order to achieve better removal efficiencies for a rather broad range of pollutants. Particular attention should be devoted to the maintenance of the reactor conserving the pellet morphology over long periods of time.

2.4.2 Removal of inorganic pollutants

The use of fungal biomass for the removal of heavy metal and metalloid ions from aqueous solution has attracted a lot of attention in the last two decades. The removal of metal ions such as AsO_3^{3-}, Cu^{2+}, Ni^{2+}, Pb^{2+}, Hg^{2+}, Cr^{3+}, $Cr_2O_7^{2-}$, CrO_4^{2-} and Zn^{2+} using fungal pellets has been reported (Kogej et al., 2010; Pakshirajan et al., 2013; Chen et al., 2011; Sepehr et al., 2012; Gabriel et al., 2001). The removal of other inorganic compounds such as selenite (SeO_3^{2-}), a toxic oxyanion of selenium, has also been treated in a fungal pelleted reactor (Espinosa-Ortiz et al., 2015b). Table 2.5 summarizes the different fungal species that have been used as pellets for the removal of heavy metals and metalloids from aqueous solution in batch experiments. The presence of numerous active sites with functional groups such as hydroxyl, carboxyl or amino in the cell walls of fungal biomass makes fungi very attractive for the biosorption of metal ions. Fungi can remove heavy metals from wastewater by different means (Tobin et al., 1994): (i) production of extracellular metabolites (*i.e.*, oxalic acid, polysaccharides and melanines), proteins and polypeptides, which are mainly produced in the presence of toxic levels of certain metals, (ii) chemical transformations (*i.e.*, redox changes or alkylation), (iii) bioaccumulation, which includes metabolism dependent processes, and (iv) biosorption, which does not involve any metabolic processes. The main mechanism for removal of metals by fungal biomass is biosorption, which can be attributed to ion exchange, coordination or covalent bonding, and adsorption, the latter being the most prominent (Barakat, 2011).

Biosorption of metals can be performed by using both living and dead fungal cells. The use of dead or inactivated fungal biomass is preferable since nutrient supply is not required and the recovery of metal species from the biomass can be easily achieved by using an appropriate desorption method (Gadd, 1990). In the case of active (living) biomass, along with the biosorption, processes that involve metabolic energy or transport might occur simultaneously. Removal of metals by fungal biomass can be enhanced when biomass is pre-treated to permeabilize the fungal cells and increase their adsorption capacity. Pre-treatments may include grinding and use of acids or alkalis, which can in turn improve the surface properties of the fungal biomass (Yetis et al., 2000).

Although there is resourceful information in the literature on the biosorption of heavy metals by living and dead fungal pellets, their use in pilot and/or full-scale reactors has been sparsely investigated. Kogej and Pavko (2001) tested the biosorption capacity of *Rhizopus nigricans* pellets to remove Pb^{2+} in a batch stirred reactor and in a packed column reactor. The authors observed a higher removal capacity of *R. nigricans* pellets in the STR (83.5 mg g^{-1}, 98%) compared to that obtained in the packed column (56 mg g^{-1}, 67%). The low performance efficiency of the continuous packed bed bioreactor was attributed to the non-uniformity packing in the bed, mainly due to the natural tendency of the pellets to form agglomerates. This could lead to the compression of the filter-bed and eventually channeling and short-circuiting of the liquid flow and decreased reactor performance. Some commonly advised strategies to prevent excess fungal growth include shortening the biological cycle and its cellular development, better process control of the physicochemical environment, and removing excess pellets.

The removal of SeO_3^{2-} (10 mg L^{-1}) in an up-flow fungal reactor using pellets of *P. chrysosporium* was first described by Espinosa-Ortiz et al. (2015b). The system showed to be suitable for the continuous removal of SeO_3^{2-} over a period of 41 d, with an average removal efficiency of 70% in the continuous operational period. In that study, different strategies to control the overgrowth of the fungal pellets were applied. For instance, mechanical removal of biomass demonstrated to be efficient; however, it also led to a periodic decrease of the biomass concentration in the system, along with a decrease in the SeO_3^{2-} removal efficiency. The system was also observed to be resilient to spikes of intermittently tested high SeO_3^{2-} concentrations. The development of fungal morphology was also tracked by the authors and it was reported that a deterioration of the pellet form was observed after 35 d of continuous operation. Therefore, the authors recommended the use of non-sterile conditions for successful long-term operation of fungal pelleted reactors, especially when dealing with toxic compounds such as SeO_3^{2-}.

Table 2.5 Biosorption of heavy metals and metalloids by fungal pellets.

Fungal species	Metal ion	pH	T (°C)	Initial metal concentration (mg L^{-1})	Metal uptake q_m (mg g^{-1})	References
Aspergillus niger	CrO_4^{2-}, Cu^{2+}, Zn^{2+}, Ni^{2+}	2.0-7.0	25	10	7.2, 5.66, 4.7, 14.1	Filipović-Kovačević et al., 2010
	Cr^{3+}	3.0-8.0	10-45	1000-1300	208.7	Sepehr et al., 2012
Lepista nuda	Cu^{2+}	3.5-4.0	25	100	6.29	Gabriel et al., 2001
Mucor indicus	Pb^{2+}	5.5	20	10	22.1	Javanbakht et al., 2011
Penicillium citrinum	Cu^{2+}	5.0	30	20	3.38	Verma et al., 2013
Phanerochaete chrysosporium	$Cr_2O_7^{2-}$	3.0	25	200	344.8	Chen et al., 2011
	AsO_3^{3-}	5.0-9.0		0.2-1		Pakshirajan et al., 2013
	Pb^{2+}	5.0	35	50-200	176.3	Xu et al., 2012
	Pb^{2+}	3.0-4.0	35	10	80	Yetis et al., 2000
	Cd^{2+}, Pb^{2+}, Cu^{2+}	2.0-7.0	25	5-500	27.7, 85.8, 26.5	Say et al., 2001
Pleurotus ostreatus	Cu^{2+}	3.5-4.0	25	100	4.77	Gabriel et al., 2001
Pycnoporus cinnabarinus	Cu^{2+}	3.5-4.0	25	100	5.08	Gabriel et al., 2001
Rhizopus nigricans	Pb^{2+}	5.5	25	10-300	83.5	Rodriguez-Couto, 2009

2.5 Scope for further research

Sterilization of wastewater is not a cost-efficient and suitable option for its treatment, and this is one of the main drawbacks of fungal reactors. However, the fungal-bacterial combination for the removal of pollutants in fungal pelleted reactors has shown, in particular cases, to even enhance the degradation of certain pollutants (Gros et al., 2014; Ferrando-Climent et al., 2015). Therefore, the development of a symbiotic fungal-bacterial consortium for the removal of pollutants from wastewater is required. Comparison of reactor configurations under different conditions is complicated and is often not accurate. Assessment of the removal of pollutants using different fungal pelleted reactor configurations under the same operational conditions and using the same wastewater to be treated would provide more accurate information on selecting the best reactor configuration to promote fungal pellet growth.

The use of fungal pellet reactors is promising for the removal of organic and inorganic pollutants from wastewater. However, their application at pilot and full-scale is still lacking. Further research is needed to better understand the potential use and operational challenges frequently encountered in full-scale applications. A recurrent issue in fungal reactors is the relatively high concentrations of COD left in the treated effluent. Further research on post-treatment technologies, as a polishing step, to remove the residual organic matter should be addressed, along with the possibility of reusing the treated effluent as irrigation water.

2.6 Conclusions

Fungal biotechnology for wastewater treatment using pelleted reactors has a wide range of applications. The use of self-immobilized fungi as pellets for wastewater treatment has shown promising results in lab-scale, achieving good success allowing the recycling of the fungal biomass and the potential recovery of products. Various types of fungi-inoculated bioreactors have been used for the removal of both organic and inorganic pollutants. Among them, the airlift reactor is the most promising reactor type to maintain the pellet shape during long term reactor operation. Further developments to full scale application of these bioreactors depend on maintaining a stable activity of the fungal pellets over prolonged periods of time as well as a good performance under non-sterile conditions.

2.7 References

Andleeb S., Atiq N., Robson G.D., Ahmed S. (2012) An investigation of anthraquinone dye biodegradation by immobilized *Aspergillus flavus* in fluidized bed bioreactor. Environ Sci Pollut Res Int 19:1728–1737.

Ajmera S.K., Delattre C., Schmidt M.A., Jensen K.F. (2002) Microfabricated cross-flow chemical reactor for catalyst testing. Sens. Actuators B Chem 82:297–306.

Babiĉ J., Pavko A. (2012) Enhanced enzyme production with the pelleted form of *D. squalens* in laboratory bioreactors using added natural lignin inducer. J Ind Microbiol Biotechnol 39:449–457.

Baccar R., Blánquez P., Bouzid J., Feki M., Attiya H., Sarrà M. (2011) Decolorization of a tannery dye: from fungal screening to bioreactor application. Biochem Eng J 56: 84–189.

Bai D.M., Jia M.Z., Zhao X.M., Ban R., Shen F., Li X.G., Xu S.M. (2003) L(+)-lactic acid production by pellet from *Rhizopus oryzae* R1021 in a stirred tank fermentor. Chem Eng Sci 58:785–791.

Bailey J.E., Ollis D.F. (1986) Biochemical Engineering Fundamentals, second ed., McGraw-Hill, New York.

Barakat M.A. (2011) New trends in removing heavy metals from industrial wastewater. Arab J Chem 4:361–377.

Bayramoğlu G., Arıca M.Y. (2008) Removal of heavy mercury (II), cadmium (II) and zinc (II) metal ions by live and heat inactivated *Lentinus edodes* pellets. Chem Eng J 143:133–140.

Bayramoğlu G., Gursel I., Tunali Y., Arika M.Y. (2009) Biosorption of phenol and 2- chlorophenol by *Funalia trogii* pellets. Bioresour Technol 100:2685–2691.

Bermek H., Guseren I., Li K., Jung H., Tamerler C. (2004) The effect of fungal morphology on ligninolytic enzyme production by a recently isolated wood degrading fungus *Trichophyton rubrum* LSK-27. World J Microbiol Biotechnol 20:345–349.

Bird R.B., Stewart W.E., Lightfoot (2002) Transport Phenomena, second ed., John Wiley and Sons Inc., New York.

Blánquez P., Guieysse B. (2008) Continuous biodegradation of 17b-estradiol and 17a-ethynnylestradiol by *Trametes versicolor*. J Hazard Mater 150:459–462.

Blánquez P., Sarra M., Vicent T. (2008) Development of a continuous process to adapt the textile wastewater treatment by fungi to industrial conditions. Proc Biochem 43:1–7.

Borchert M., Libra J.A. (2001) Decolorization of reactive dyes by the white rot fungus *Trametes versicolor* in sequencing batch reactors. Biotechnol Bioeng 75:313–321.

Bosso L., Lacatena F., Cristinzio G., Cea M., Diez M.C., Rubilar O. (2015) Biosorption of pentachlorophenol by *Anthracophyllum discolor* in the form of live fungal pellets. New Biotechnol 32:21–25.

Braun S., Vecht-Lifsitz S.E. (1991) Mycelial morphology, and metabolite production. Trends Biotechnol 9:63–68.

Cabaleiro D.R., Rodriguez-Couto S., Sanromán A., Longo M.A. (2002) Comparison between the protease production ability of ligninolytic fungi cultivated in solid state media. Process Biochem 37:1017–1023.

Calam C.T. (1976) Starting investigational and production cultures. Process Biochem 11:7–12.

Callow N.V., Ju L.K. (2012) Promoting pellet growth of *Trichoderma reesei* Rut C30 by surfactants for easy separation and enhanced cellulase production. Enzyme Microb Technol 50:311–317.

Casas-López J.L., Sánchez-Pérez J.A., Fernández-Sevilla J.M, Acién Fernández F.G., Molina Grima E., Chisti Y. (2003) Production of lovastatin by *Aspergillus terreus*: effects of the C:N ratio and the principal nutrients on growth and metabolite production. Enzyme Microb Technol 33:270–277.

Cao J., Zhang H.J., Xu C.P. (2014) Culture characterization of exopolysaccharides with antioxidant activity produced by *Pycnoporus sanguineus* in stirred-tank and airlift reactors. J Taiwan Institute Chem Eng 45:2075–2080.

Cascaval D., Galaction A.I., Oniscu C., Ungureanu F. (2004) Modeling of mixing in stirred bioreactors 4. Mixing time for aerated bacteria, yeasts and fungus broths. Chem Ind 59:128–137.

Cerrone F., Barghini P., Pesciaroli C., Fenice M. (2011) Efficient removal of pollutants from olive washing wastewater in bubble-column bioreactor by *Trametes versicolor*. Chemosphere 84: 254–259.

Chen G.Q., Zhang W.J., Zeng G.M., Huang J.H., Wang L., Shen G.L. (2011) Surface modified *Phanerochaete chrysosporium* as a biosorbent for Cr(VI)- contaminated wastewater. J Hazard Mater 186:2138–2143.

Cheng Z., Xiang-Hua W., Ping N. (2013) Continuous Acid Blue 45 decolorization by using a novel open fungal reactor system with ozone as the bactericide. Biochem Eng J 79:246–252.

Cho Y.J., Hwang H.J., Kin S.W., Song C.H., Yun J.W. (2002) Effect of carbon source and aeration rate on broth rheology and fungal morphology during red pigment production by *Paecilomyces sinclairii* in a batch bioreactor. J Biotechnol 95:13–23.

Colin V.L., Baigorí M.D., Pera L.M. (2013) Tailoring fungal morphology of *Aspergillus niger* MYA 135 by altering the hyphal morphology and the conidia adhesion capacity: biotechnological applications. AMB Express 3:27.

Cruz-Morató C., Lucas D., Llorca M., Rodriguez-Mozaz S., Gorga M., Petrovic M., Barceló D., Vicent T., Sarrá M., Marco-Urrea E. (2014) Hospital wastewater treatment by fungal bioreactor: removal efficiency for pharmaceuticals and endocrine disruptor compounds. Sci Total Environ 493:356–376.

Cruz-Morató C., Ferrando-Climent L., Rodriguez-Mozaz S., Barceló D., Marco-Urrea E., Vicent T., Sarrà M. (2013) Degradation of pharmaceuticals in nonsterile urban wastewater by *Trametes versicolor* in a fluidized bed bioreactor. Water Res 47:5200–5210.

Cui Y.Q., Okkerse W.J., van der Lans R.G.J.M., Luyben K. (1998) Modeling and measurements of fungal growth and morphology in submerged fermentations. Biotechnol Bioeng 60:216–229.

Domingues F.C., Queiroz J.A., Cabral J.M.S., Fonseca L.P. (2000) The influence of culture conditions on mycelial structure and cellulase production by *Trichoderma reesei* Rut C-30. Enzyme Microb Technol 26:394–401.

Grimm L.H., Kelly S., Krull R., Hempel D.C. (2005) Morphology and productivity of filamentous fungi. Appl Microbiol Biotechnol 69:375–384.

El-Enshasy H.A. (2007) Filamentous fungal cultures – process characteristics, products, and applications, in: S.T. Yang (Ed.), Bioprocessing for Value-Added Products from Renewable Resources – New Technologies and Applications, Elsevier, Amsterdam, pp. 225–261.

Espinosa-Ortiz E.J., Gonzalez-Gil G., Salikaly P.E., van Hullebusch E.D., Lens P.N.L. (2015a) Effects of selenium oxyanions on the white-rot fungus *Phanerochaete chrysosporium*. Appl Microbiol Biotechnol 99:2405–2418.

Espinosa-Ortiz E.J., Rene E.R., van Hullebusch E.D., Lens P.N.L. (2015b) Removal of selenite from wastewater in a *Phanerochaete chrysosporium* pellet based fungal bioreactor. Int Biodeterior Biodegrad 102:361–369.

Farkas V., Felinger A., Hegedüsova A., Dékány I., Pernyeszi T. (2013) Comparative study of the kinetics and equilibrium of phenol biosorption on immobilized white-rot fungus *Phanerochaete chrysosporium* from aqueous solution. Colloid Surf B Biointerfaces 103:381–390.

Fassnacht D., Portner R. (1999) Experimental and theoretical considerations on oxygen supply for animal cell growth in fixed-bed reactors. J Biotechnol 72:169–184.

Ferrando-Climent L., Cruz-Morató C., Marco-Urrea E., Vicent T., Sarra M., Rodríguez-Mozaz S., Barceló D. (2015) Non conventional biological treatment based on *Trametes versicolor* for the elimination of recalcitrant anticancer drugs in hospital wastewater. Chemosphere 136: 9–19.

Filipović-Kovačević Z., Sipos L., Briški F. (2010) Biosorption of chromium, copper, nickel, and zinc ions onto fungal pellets of *Aspergillus niger* 405 from aqueous solutions. Food Technol Biotechnol 38:211–216.

Font G., Caminal X., Gabarrell S., Romero M.T., Vicent. (2003) Black liquor detoxification by laccase of *Trametes versicolor* pellets. J Chem Technol Biotechnol 78:548–554.

Font X., Caminal G., Gabarell X., Vicent T. (2006) Treatment of toxic industrial wastewater in fluidized and fixed-bed batch reactors with *Trametes versicolor*: influence of immobilization. Environment Technol 27: 845–854.

Fontana R.C., da Silveira M.M. (2012) Production of polygalacturonases by *Aspergillus oryzae* in stirred tank and internal- and external-loop airlift reactor, Bioresour Technol 123: 157–163.

Fu Y., Viraraghavan T. (2001) Fungal decolorization of dye wastewaters: a review. Bioresour Technol 79:251–262.

Gadd G.M. (1990) Fungi and yeasts for metal binding, in: H.L. Ehrlich, C.L. Brierley (Eds.), Microbial Mineral Recovery, McGraw-Hill, New York, pp. 249–275.

Gao D., Zeng Y., Wen X., Qian Y. (2008) Competition strategies for the incubation of white rot fungi under non-sterile conditions. Process Biochem 43:937–944.

García-Soto M.J., Botello-Álvarez E., Jiménez-Islas H., Navarrete-Bolaños J., Barajas-Conde E., Rico-Martínez R. (2006) Growth morphology and hydrodynamics of filamentous fungi in submerged cultures. Adv Agricultural Food Biotechnol 17–34.

Glicksman L.R. (1999) Fluidized bed scale-up, in: W.C. Yang (Ed.), Fluidization, Solids Handing and Processing, Noyes Publications, New Jersey, USA.

Golan-Rozen N., Chefetz B., Ben-Ari J., Geva J., Hadar Y. (2011) Transformation of the recalcitrant pharmaceutical compound carbamazepine by *Pleurotus ostreatus*: role of cytochrome P450 monooxygenase and manganese peroxidase. Environ Sci Technol 45:6800–6805.

Grimm L.H., Kelly S., Volkerding I.I., Krull R., Hempel D.C. (2005) Influence of mechanical stress and surface interaction on the aggregation of *Aspergillus niger* conidia. Biotechnol Bioeng 92:879–888.

Gros M., Cruz-Mortao C., Marco-Urrea E., Longrée P., Singer H., Sarrá M., Hollender J., Vicent T., Rodriguez-Mozaz S., Barceló D. (2014) Biodegradation of the X-ray contrast agent iopromide and the fluoroquinolone antibiotic ofloxacin by the white rot fungus *Trametes versicolor* in hospital wastewaters and identification of degradation products. Water Res 60:228–241.

Hai F.I., Yamamoto K., Nakajima F., Fukushi K., Nghiem L.D., Price W.E., Jin B. (2013) Degradation of azo dye Acid Orange 7 in a membrane bioreactor by pellets and attached growth of *Coriolus versicolour*. Bioresour Technol 41:29–34.

Harms H., Schlosser D., Wick L.Y. (2011) Untapped potential: exploiting fungi in bioremediation of hazardous chemicals. Nat Rev Microbiol 9:177–192.

Higashiyama K.K., Murakami K., Tsujimura H., Matsumoto N., Fujikawa S. (1999) Effects of dissolved oxygen on the morphology of an arachidonic acid production by *Mortierella alpine* 1S–4. Biotechnol Bioeng 63:442–448.

Jaganathan B., Masud-Hossain S.K., Meera Sheriffa Begum K.M., Anantharaman N. (2009) Aerobic pollution abatement of pulp mill effluent with the white rot fungus *Phanerochaete chrysosporium* in three-phase fluidized bed bioreactor. Chem Eng Res Bull 13–16.

Javanbakht V., Zilouei H., Karimi K. (2011) Lead biosorption by different morphologies of fungus *Mucor indicus*. Int Biodeterior Biodegrad 65:294–300.

Jelic A., Cruz-Morató C., Marco-Urrea E., Sarrà M., Perez S., Vicent T., Petrovic M., Barcelo D. (2012) Degradation of carbamazepine by *Trametes versicolor* in an air pulsed fluidized bed bioreactor and identification of intermediates. Water Res 46:955–964.

Jesús A.G., Romano-Baez F.J., Leyva-Amezcua L., Juárez-Ramírez C., Ruiz-Ordaz N., Galíndez-Maye J. (2009) Biodegradation of 2,4,6-trichlorophenol in a packed bed biofilm reactor equipped with an internal net draft tube riser for aeration and liquid circulation. J Hazard Mater 61:1140–1149.

Jeyakumar D., Christeen J., Doble M. (2013) Synergistic effects of pretreatment and blending on fungi mediated biodegradation of polypropylenes. Bioresour Technol 148:78–85.

Jonsbu E., McIntyre M., Nielsen J. (2002) The influence of carbon sources and morphology on nystatin production by *Strepomyces noursei*. J Biotechnol 95:133–144.

Jordan M., Sucker H., Einsele A., Widmer F., Eppenberger H.M. (1994) Interactions between animal cells and gas bubbles: the influence of serum and pluronic F68 on the physical properties of the bubble surface. Biotechnol Bioeng 43:446–454.

Kanai T., Uzumaki T., Kawase Y. (1996) Simulation of airlift bioreactors: steady-state performance of continuous culture processes. Comp Chem Eng 20:1089–1099.

Kandelbauer A., Maute O., Kessler R.W., Erlacher A., Gübitz G.M. (2004) Study of dye decolourization in an immobilized laccase enzyme-reactor using online spectroscopy. Biotechnol Bioeng 87:552–563.

Kantarci N., Borak F., Ulgen K.O. (2005) Bubble column reactors. Proc Biochem 40: 2263–2283.

Kaushik P., Mishra A., Malik A., Pant K.K. (2014) Biosorption of textile dye by *Aspergillus lentulus* pellets: process optimization and cyclic removal in aerated bioreactor. Water Air Soil Pollut 225:1978.

Kennes C., Veiga M.C. (2004) Fungal biocatalysts in the biofiltration of VOC-polluted air. J Biotechnol 113:305–319.

Kim Y., Song H. (2009) Effect of fungal pellet morphology on enzyme activities involved in phthalate degradation. J Microbiol 47:420–424.

Kogej A., Pavko A. (2001) Laboratory experiments of lead biosorption by selfimmobilized *Rhizopus nigricans* pellets in the batch stirred tank reactor and the packed bed column. Chem Biochem Eng Q 15:75–79.

Kogej A., Likozar B., Pavko A. (2010) Lead biosorption by self-immobilized *Rhizopus nigricans* pellets in a laboratory scale packed bed column: mathematical model and experiment. Food Technol Biotechnol 48:344–351.

Kreiner M., Harvey L.M., McNeil B. (2003) Morphological and enzymatic responses of a recombinant *Aspergillus niger* to oxidative stressors in chemostat cultures. J Biotechnol 100:251–260.

Kretzmer G. (2002) Industrial processes with animal cells, Appl Microbiol Biotechnol 59:135–142.

Kunii D., Levenspiel O. (1968) Bubbling bed model for kinetic processes in fluidized beds. Ind Eng Chem Process Des Dev 7:481–492.

Li X., Alves de Toledo R., Wang S., Shim H. (2015) Removal of carbamazepine and naproxen by immobilized *Phanerochaete chrysosporium* under non-sterile condition. New Biotechnol 32:282–289.

Lema J.M., Roca E., Sanromán A., Núñez M.J., Moreira M.T., Feijoo G. (2001) Pulsing bioreactors. In: J.M.S. Cabral, M. Mota, J. Tramper (Eds.), Multiphase Bioreactor Design, Taylor & Francis, London, pp. 309–329.

Liao W., Liu Y., Frear C., Chen S. (2007) A new approach of pellet formation of a filamentous fungus – *Rhizopus oryzae*. Bioresour Technol 98:3415–3423.

Libra J.A., Borchert M., Banit S. (2003) Competition strategies for the decolorization of a textile-reactive dye with the white-rot fungi *Trametes versicolor* under non-sterile conditions. Biotechnol Bioeng 82:736–744.

Marco-Urrea E., Pérez-Trujillo M., Cruz-Morató C., G. Caminal, T. Vicent, White-rot fungus-mediated degradation of the analgesic ketoprofen and identification of intermediates by HPLC-DAD-MS and NMR. Chemosphere 78:474–481.

Merchuk J.C., Gluz M. (1991) Bioreactors, air-lift reactors, in: M.C. Flickinger, S.W. Drew (Eds.), Encyclopedia of Bioprocess Technology, John Wiley & Sons, New York, pp. 320–350.

Metz B., Kossen N.W.F. (1977) The growth of molds in the form of pellets – a literature review. Biotechnol Bioeng 19:781–799.

Metz B., Kossen N.W.F., van Suijdam J.C. (1979) The rheology of mold suspensions. Adv Biochem Eng 11:103–156.

Mir-Tutusaus J.A., Masis-Mora M., Corcellas C., Eljarrat E., Barceló D., Sarrá M., Caminal G., Vicent T., Rodríguez-Rodríguez C.E. (2014) Degradation of selected agrochemicals by the white-rot fungus *Trametes versicolor*. Sci Total Environ 500–501:235–242.

Moreira M.T., Feijoo G., Lema J.M. (2003) Fungal bioreactors: applications to white-rot fungi. Rev Environ Sci Biotechnol 2:247–259.

Moreira M.T., Sanromán A., Feijoo G., Lema J.M. (1996) Control of pellet morphology of filamentous fungi in fluidized bed bioreactors by means of a pulsing flow. Application to *Aspergillus niger* and *Phanerochaete chrysosporium*. Enzyme Microb Technol 19:261–266.

Murugesan K., Kalaichelvan P.T. (2003) Synthetic dye decolourization by white rot fungi. Indian J Exp Biol 41:1076–1087.

Nanou K., Roukas T., Papadakis E. (2012) Improved production of carotenes from synthetic medium by *Blakeslea trispora* in a bubble column reactor. Biochem Eng J 67:203–207.

Nanou K., Roukas T., Papadakis E. (2011) Oxidative stress and morphological changes in *Blakeslea trispora* induced by enhanced aeration during carotene production in a bubble column reactor. Biochem Eng J 54:172–177.

Nielsen J., Villadsen J. (1994) Bioreaction Engineering Principles, Plenum, New York, USA.

Nielsen J. (1996) Modelling the morphology of filamentous microorganisms. Trends Biotechnol 14:438–443.

Núñez-Ramirez D.M., Valencia Lopez J.J., Calderas F., Solis-Soto A., Lopez-Miranda J., Medrano-Rodlan H., Medina-Torres L. (2012) Mixing analysis for a fermentation broth of the fungus *Beauveria bassiana* under different hydrodynamic conditions in a bioreactor. Chem Eng Technol 35:1954–1961.

Ntougias S., Baldrian P., Ehaliotis C., Nerud F., Merhautová V., Zervakis G.I. (2015) Olive mill wastewater biodegradation potential of white rot fungi – mode of action of fungal culture extracts and effect of ligninolytic enzymes. Bioresour Technol 189:121–130.

Pakshirajan K., Swaminathan T. (2009) Biosorption of copper and cadmium in packed bed columns with live immobilized fungal biomass of *Phanerochaete chrysosporium*. Appl Biochem Biotechnol 157:159–173.

Pakshirajan K., Izquierdo M., Lens P.N.L. (2013) Arsenic (III) removal at low concentrations by biosorption using *Phanerochaete chrysosporium* pellets. Sep Sci Technol 48: 1111–1122.

Papagianni M. (2007) Advances in citric acid fermentation by *Aspergillus niger*: biochemical aspects, membrane transport and modeling. Biotechnol Adv 25: 244–263.

Papagianni M. (2004) Fungal morphology and metabolite production in submerged mycelial processes. Biotechnol Adv 22:189–259.

Papagianni M., Moo-Young M. (2002) Protease secretion in glucoamylase produces *Aspergillus niger* cultures: fungal morphology and inoculum effects. Process Biochem 37: 1271–1278.

Papoutsakis E.T. (1991) Media additives for protecting freely suspended animal cells against agitation and aeration damage. Trends Biotechnol 9:316–324.

Paul G.C., Thomas C.R. (1998) Characterization of mycelial morphology using image analysis. Adv Biochem Eng Biotechnol 60:2–59.

Pazouki M., Panda T. (2000) Understanding the morphology of fungi. Bioprocess Eng 22:127–143.

Réczey K., Stalbrand H., Hahn-Hagerdal B., Tjerneld F. (1992) Mycelia-associated β-galactosidase activity in microbial pellets of *Aspergillus* and *Penicillium strains*. Appl Microbiol Biotechnol 38: 393–397.

Rene E.R., Veiga M.C., Kennes C. (2010) Biodegradation of gas-phase styrene using the fungus *Sporothrix variecibatus*: Impact of pollutant load and transient operation. Chemosphere 79:221–227.

Renganathan S., Thilagaraj W.R., Miranda L.R., Gautam P., Velan M. (2206) Accumulation of Acid Orange 7, Acid Red 18 and Reactive Black 5 by growing *Schizophyllum commune*. Bioresour Technol 97:2189–2193.

Rioja R., García M.T., Peña M., González G. (2008) Biological decolourisation of wastewater from molasses fermentation by *Trametes versicolor* in an airlift reactor. J Environ Sci Health A Tox Hazard Subst Environ Eng 43:772–778.

Rodríguez-Rodríguez C.E., García-Galán M.J., Blánquez P., Díaz-Cruz M.S., Barceló D., Caminal G., Vicent T. (2012) Continuous degradation of a mixture of sulfonamides by *Trametes versicolor* and identification of metabolites from sulfapyridine and sulfathiazole. J Hazard Mat 213–214: 347–354.

Rodríguez-Couto S. (2012) A promising inert support for laccase production and decolouration of textile wastewater by the white-rot fungus *Trametes pubescesns*. J Hazard Mater 223–234: 158–162.

Rodriguez-Couto S. (2009) Dye removal by immobilised fungi. Biotechnol Adv 27:227–235.

Rodríguez-Porcel E.M., Casas-López J.L., Sánchez-Pérez J.A., Fernández- Sevilla J.M., Chisti Y. (2005) Effects of pellet morphology on broth rheology in fermentations of *Aspergillus terreus*. Biochem Eng J 26:139–144.

Rodríguez-Porcel E.M., Casas-López J.L., Sánchez-Pérez J.A., Chisti Y. (2007) Enhanced production of lovastatin in a bubble column by *Aspergillus terreus* using a two-stage feeding strategy. J Chem Technol Biotechnol 82:58–64.

Rodarte-Morales A.I., Feijoo G., Moreira M.T., Lema J.M. (2012) Biotransformation of three pharmaceutical active compounds by the fungus *Phanerochaete chrysosporium* in a fed batch stirred reactor under air and oxygen supply. Biodegradation 23:145–156.

Rodarte-Morales A.I., Feijoo G., Moreira M.T., Lema J.M. (2012) Operation of stirred tank reactors (STRs) and fixed-bed reactors (FBR's) with free and immobilized *Phanerochaete chrysosporium* for the continuous pharmaceutical compounds. Biochem Eng J 66:38–45.

Rubilar O., Elgueta S., Tortella G., Gianfreda L., Diez M.C. (2009) Pelletization of *Anthracophyllum discolor* for water and soil treatment contaminated with organic pollutants. J Soil Sci Plant Nutr 9:161–175.

Ryan D.R., Leukes W.D., Burton S.G. (2005) Fungal bioremediation of phenolics wastewaters in an airlift reactor. Biotechnol Progr 21:1066–1074.

Sanghi R., Dixit A., Guha S. (2006) Sequential batch culture studies for the decolorization of reactive dye by *Coriolus versicolor*. Bioresour Technol 97:396–400.

Sankaran S., Khanal S.K., Pomettro III A.L., van Leeuwen J. (2008) Ozone as a selective disinfectant for nonaseptic fungal cultivation on corn-processing wastewater. Bioresour Technol 99:8265–8272.

Saraswathy A., Hallberg R. (2005) Mycelial pellet formation *by Penicillium ochrochloron* species due to exposure to pyrene. Microbiol Res 160:375–383.

Say R., Denizli A., Arica M.Y. (2001) Biosorption of cadmium(II), lead(II) and copper(II) with the filamentous fungus *Phanerochaete chrysosporium*. Bioresour Technol 76:67–70.

Sepehr M.N., Nasseri S., Zarrabi M., Samarghandi M.R., Amrane A. (2012) Removal of Cr(III) from tanning effluent by *Aspergillus niger* in airlift bioreactor. Sep Purif Technol 9:256–262.

Sharma A., Padwal-Desai S.R. (1985) On the relationship between pellet size and aflatoxin yield in *Aspergillus parasiticus*. Biotechnol Bioeng 27:1577–1580.

Sumathi S., Manju B.S. (2000) Uptake of reactive textile dyes by *Aspergillus foetidus*. Enzyme Microb Technol 27:347–355.

Sulaiman S.S., Annuar B.S.M., Razak N.N.A., Ibrahim S., Bakar B. (2013) Triatylmethane dye decolorization by pellets of *Pycnoporus sanguineus*: statistical optimization and effects of novel impeller geometry. Bioremediat J 17:305–315.

Taseli B.K., Gökçay C.F., Taeli H. (2004) Upflow column reactor design for dechlorination of chlorinated pulping wastes *Penicillium camemberti*. J Environ Manage 72:175–179.

Tobin J.M., White C., Gadd G.M. (1994) Metal accumulation by fungi applications in environmental biotechnology. J Ind Microbiol 13:126–130.

Thongchul N., Yang S.T. (2003) Controlling filamentous fungal morphology by immobilization on a rotating fibrous matrix to enhance oxygen transfer and L(+)-lactic acid production by *Rhizopus oryzae*, in: B.C. Saha (Ed.), ACS Symposium Series 862, Fermentation Process Development, Oxford University Press, New York, pp. 36–51.

van Leeuwen J.H., Hu Z., Yi T., Pometto A.L.I.I.I., Jin B. (2003) Kinetic model for selective cultivation of microfungi in a microscreen process for food processing wastewater treatment and biomass production. Acta Biotechnol 23:289–300.

van der Westhuizen T.H., Pretorius W.A. (1998) Use of filamentous fungi for the purification of industrial effluents, WRC Report No. 535/I/98, Water Research Commission, Pretoria, South Africa.

Verma A., Shalu A., Singh N.R., Bishnoi A., Gupta (2013) Biosorption of Cu(II) using free and immobilized biomass of *Penicillium citrinum*. Ecol Eng 61A:486–490.

Wang S.J., Zhong J.J. (2007) Bioreactor Engineering, in: S.T. Yang (Ed.), Bioprocessing for Value-Added Products from Renewable Resources, Elsevier, Amsterdam, pp. 131–161.

Wang M.X., Zhang Q.L., Yao S.J. (2015) A novel biosorbent formed of marine-derived *Penicillium janthinellum* mycelial pellets for removing dyes from dye containing wastewater. Chem Eng J 259:837–844.

Wittier R., Baumgartl H., Lübbers D.W., Schügerl K. (1986) Investigations of oxygen transfer into *Penicillium chrysogenum* pellets by microprobe measurements. Biotechnol Bioeng 28:1024–1036.

Wucherpfennig T., Hestler T., Krull R. (2011) Morphology engineering - Osmolality and its effect on *Aspergillus niger* morphology and productivity. Microb Cell Fact 10:58–73.

Xin B., Xia Y., Zhang Y., Aslam H., Liu C., Chen S. (2012) A feasible method for growing fungal pellets in a column reactor inoculated with mycelium fragments and their application for dye bioaccumulation from aqueous solution. Bioresour Technol 105:100–105.

Xin B., Chen G., Zheng W. (2010) Bioaccumulation of Cu-complex reactive dye by growing pellets of *Penicillium oxalicum* and its mechanism. Water Res 44:3565–3572.

Xin B., Zhang Y., Liu C., Chen S., Wu B. (2012) Comparison of specific adsorption capacity of different forms of fungal pellets for removal of Acid Brilliant Red B from aqueous solution and mechanisms exploration. Process Biochem 47:1197–1201.

Xu P., Zeng G.M., Huang D.L., Lai C., Zhao M.H., Wei Z., Li N.J., Huang C., Xie G.X. (2012) Adsorption of Pb(II) by iron oxide nanoparticles immobilized *Phanerochaete chrysosporium*: equilibrium, kinetic, thermodynamic and mechanisms analysis. Chem Eng J 203:423–431.

Yang S., Hai F.I., Nghiem L.D., Nguyen L.N., Roddick F., Price W.E. (2013) Removal of bisphenol A and diclofenac by a novel fungal membrane bioreactor operated under non-sterile conditions. Int Biodeterior Biodegrad 85:483–490.

Yang S., Hai F.I., Nghiem L.D., Price W.E., Roddick F., Moreira M.T., Magram S.F. (2013) Understanding the factors controlling the removal of trace organic contaminants by white-rot fungi and their lignin modifying enzymes: a critical review. Bioresour Technol 141:97–108.

Yesilada O., Cing S., Asma D. (2002) Decolourisation of the textile dye Astrazon Red FBL by *Funalia trogii* pellets. Bioresour Technol 81:155–157.

Yetis U., Dolek A., Dilek F.B., Ozcengiz G. (2000) The removal of Pb(II) by *Phanerochaete chrysosporium*. Water Res 34:4090–4100.

Yu L., Chao Y., Wensel P., Chen S. (2012) Hydrodynamic and kinetic study of cellulase production by *Trichoderma reesei* with pellet morphology. Biotechnol Bioeng 109:1755–1768.

Zhang Y., Arends Y.B.A., de Wiele T.V., Boon N. (2011) Bioreactor technology in marine microbiology: from design to future application. Biotechnol Adv 29:312–321.

Zhang S., Li A., Cui D., Yang J., Ma F. (2011) Performance of enhanced biological SBR process for aniline treatment by mycelial pellet as biomass carrier. Bioresour Technol 102:4360–4365.

Zhang Y., Geissen S.U. (2010) Prediction of carbamazepine in sewage treatment plant effluents and its implications for control strategies of pharmaceutical aquatic contamination. Chemosphere 80:1345–1352.

Zhang Z.Y., Jin B., Kelly J.M. (2007) Production of lactic acid and byproducts from waste potato starch by *Rhizopus arrhizus*: role of nitrogen sources, World J Microbiol Biotechnol 23:229–236.

Zhang F., Yu J. (2000) Decolourisation of Acid Violet 7 with complex pellets of white rot fungus and activated carbon. Bioproc Eng 2:295–301.

Zhang F.M., Knapp J.S., Tapley K.N. (1999) Development of bioreactor systems for decolourization of Orange II using white rot fungus. Enzyme Microb Technol 24:48–53.

Zhao L.H., Zhou J.T., Lv H., Zheng C.L., Yang Y.S., Sun H.J, Zhang X.H. (2008) Decoloration of cotton pulp black liquor by *Pleurotus ostreatus* in a bubble column reactor. Bulletin Environ Contamin Toxicol 80:44–48.

Zhong J.J. (2010) Recent advances in bioreactor engineering, Korean J. Chem. Eng. 27: 1035–1041.

Zhou T.C., Zhou W.W., Hu W.W., Zhong J.J. (2010) Cell culture, bioreactors, commercial production, In: M.C. Flickinger (Ed.), Encyclopedia of Industrial Biotechnology 2, Wiley, New York, pp. 913–939.

CHAPTER 3

Effects of selenium oxyanions on the white-rot fungus *Phanerochaete chrysosporium*

A modified version of this chapter was published as:

E.J. Espinosa-Ortiz, G. Gonzalez-Gil, P.E. Saikaly, E.D. van Hullebusch, P.N.L. Lens, (2015). Effects of selenium oxyanions on the white-rot fungus *Phanerochaete chrysosporium*. Applied Microbiology and Biotechnology. 99:2405–2418.

Abstract

The ability of *Phanerochaete chrysosporium* to reduce the oxidized forms of selenium, selenate and selenite, and their effects on the growth, substrate consumption rate and pellet morphology of the fungus was assessed. The effect of different operational parameters (pH, glucose and selenium concentration) on the response of *P. chrysosporium* to selenium oxyanions was explored as well. This fungal species showed a high sensitivity to selenium, particularly selenite, which inhibited the fungal growth and substrate consumption when supplied at 10 mg L^{-1} in the growth medium; whereas selenate did not have such a strong influence on the fungus. Biological removal of selenite was achieved under semi-acidic conditions (pH 4.5) with about 40% removal efficiency, whereas less than 10% selenium removal was achieved for incubations with selenate. *P. chrysosporium* was found to be a selenium-reducing organism, capable of synthesizing elemental selenium from selenite, but not from selenate. Analysis with transmission electron microscopy, electron energy-loss spectroscopy and a 3D reconstruction showed that elemental selenium was produced intracellularly as nanoparticles in the range of 30-400 nm. Furthermore, selenite influenced the pellet morphology of *P. chrysosporium* by reducing the size of the fungal pellets and inducing their compaction and smoothness.

Key words: Fungal pellets, selenium removal, selenium nanoparticles, *Phanerochaete chrysosporium*

3.1 Introduction

Selenium (Se) is well known for being both essential and toxic to human and animal life. As an essential biological trace element, the absence or deficiency in the human diet might cause major health issues (Rayman, 2012). On the other hand, the toxicity of Se is only about one order of magnitude above its essential level (Fordyce, 2013), leading to severe consequences for human health, from hair loss to dermal, respiratory and neurological damages (USHHS, 2003).

As a bulk material, Se has been used in a broad range of different applications, such as fertilizer, dietary supplement and fungicide, as well as in the glass and electronic industries. The extensive use and consumption of Se has raised plenty of concerns, since it was identified as a major threat for aquatic ecosystems in the 1980s (Hamilton, 2004). Concentrations from a few to thousand of μg Se L^{-1} of the toxic, water soluble oxyanions of Se, selenate (SeO_4^{2-}) and selenite (SeO_3^{2-}), are mainly found in the wastewater of industries associated to the production of glass, pigments, solar batteries and semiconductors, mining, refinery, coal combustion, petrochemical, thermal power stations, electronic and agricultural activities (Lenz and Lens, 2009; Soda and Ike, 2011).

Traditionally, physicochemical methods have been used for the removal of Se oxyanions from water (Bleiman and Mishael, 2010; Frankenberger et al., 2004; Geoffroy and Demopoulos, 2011; Mavrov et al., 2006; Nguyen et al., 2005; Zelmanov and Semiat, 2013). The use of biological processes is an attractive alternative, due to the ability of biological agents to reduce toxic Se species to elemental selenium (Se^0), a more stable and less toxic form of this element (Zhang et al., 2005; 2008). The synthesis of Se^0 in biological treatment systems is an added value, particularly when it is produced and recovered at the nano size range (<100 nm). Elemental selenium nanoparticles (nSe^0) possess unique semiconducting, photoelectric and X-ray sensing properties, which makes them attractive for their use in photocells, semiconductor rectifiers and photocopy machines, as well as antifungal, anti-cancer (Ahmad et al., 2010; van Cutsem et al., 1990) and therapeutic agents (Beheshti et al., 2013).

The synthesis of nSe^0 from Se oxyanions has been successfully achieved using bacteria (Zhang et al., 2011), fungi (Sarkar et al., 2011; Vetchinkina et al., 2013) and plant extracts (Mittal et al., 2013; Prasad et al., 2013). Even though bacteria are the preferred microorganisms for the production of nanoparticles, the use of fungi is also promising, due to the ease of handling and scale-up, as well as their ability to produce large amounts of enzymes and ability to survive at low pH. The capacity of fungi to act as nanofactories has already been shown for more than 30 different fungal species, which were able to synthesize diverse nanoparticles, including nAg, nAu, $nTiO_2$, and nPt (Castro-Longoria, 2011; Rajakumar et al., 2012; Syed and Ahmad, 2012; Verma et al., 2010). Although some fungal strains have been found to be Se-reducing organisms, the synthesis of nSe^0 has only been demonstrated by two fungal species, *Alternaria alternata* (Sarkar et al., 2011) and *Lentinula edodes* (Vetchinkina et al., 2013).

There has been some research on the metabolism of Se in filamentous fungi, including uptake and volatilization (Barkes and Fleming, 1974; Fleming and Alexander, 1972; Gharieb et al., 1995; Tweedie and Segel, 1970; Ramadan et al., 1988); however, the information is still scarce. The limited knowledge regarding Se-fungi interactions hampers the full exploitation of fungi in Se remediation technologies and as nSe^0 producing agents. Further investigations are required to elucidate the influence of Se on fungal morphology, growth and ability to synthesize nSe^0.

White-rot fungi are well known for their unique ability to generate ligninolytic enzymes (*i.e.,* lignin peroxidase, manganese peroxidase, laccase), which makes them attractive for the degradation of a wide range of hazardous compounds (Cameron et al., 2000; Lee et al., 2014), as well as for the synthesis of different nanoparticles (*e.g.,* nAg, Chan and Don, 2013; Vigneshwaran et al., 2006). The capacity of the white-rot fungus *Phanerochaete chrysosporium* to degrade organic compounds has been demonstrated (Moldes et al., 2003; Wang et al., 2009), as well as its ability to synthesize metal (Ag and Au) nanoparticles (Sanghi et al., 2011; Vigneshwaran et al., 2006). This fungus is able to self-immobilize in the form of pellets, which has many practical advantages over dispersed mycelial growth (Pazouki and Panda, 2000; Thongchul and Yang, 2003).

The aim of this study was to explore the potential of *P. chrysosporium* as a Se-reducing organism to treat effluents polluted with Se oxyanions and to enlarge the scope of biological agents for the mycosynthesis of nanomaterials. Specifically, the influence of glucose concentration, pH and Se oxyanion concentration on the biomass growth, pelletization and Se removal efficiency was investigated. Transmission electron microscopy (TEM) images and 3D reconstruction were used to localize the synthesized nSe^0.

3.2 Materials and methods

3.2.1 Fungal culture and medium composition

The fungal strain *P. chrysosporium* (MTCC 787) was obtained from the Institute of Microbial Technology (IMTECH), Chandigarh (India). The fungus was grown on malt agar plates at 37°C for 3 d. Subcultures were routinely prepared as required and were maintained at 4°C. A fungal spore solution was prepared by harvesting all the fungal spores from one of the 3 d old agar plates into 50 mL of nitrogen-limited liquid medium. Fungi were grown in 100 mL Erlenmeyer flasks with 50 mL of medium. The flasks were inoculated with 2% (v/v) of the spore suspension and incubated at 30°C on a rotating orbital incubator shaker (Innova 2100, New Brunswich Scientific) set at 150 rpm. Subcultures of the 2 d old fungal growth in the flasks were used as the final inoculum (2 % v/v, 0.003 g dry biomass L^{-1}). The use of a 2 day fungal growth as inoculum instead of a spore solution was found to be more efficient in terms of maintaining a reproducible formation and shape of fungal pellets. Furthermore, this approach allowed quantifying, by dry weight, the amount of biomass used as inoculum for the experiments.

The nitrogen-limited liquid medium was composed of (g L^{-1}): glucose, 10; KH_2PO_4, 2; $MgSO_4{\cdot}7H_2O$, 0.5; NH_4Cl, 0.1; $CaCl_2{\cdot}2H_2O$, 0.1; thiamine, 0.001 and 5 mL of trace elements solution (Tien and Kirk, 1988). After adjusting the pH to 4.5, the medium was sterilized at 123 kPa and 110°C for 30 min and cooled to room temperature before use.

3.2.2 Batch experiments

3.2.2.1 Fungal interaction with selenate and selenite

The effect of SeO_4^{2-} and SeO_3^{2-} on the morphology and growth of *P. chrysosporium* was assessed in batch incubations. Fungal pellets were formed and grown in the presence of either Na_2SeO_4 or

Na_2SeO_3 (10 mg Se L^{-1}) for 12 d. Abiotic controls without biomass and with dead biomass (autoclaved at 121 °C, 123 kPa) were included to account for potential abiotic reactions and potential adsorption of Se to inert biomass, respectively. Biotic controls, in which the fungus was grown in the absence of Se, were used to record growth characteristics under non-stressful conditions. In order to avoid contamination and to maintain axenic growth of the fungus, experimental flasks were sacrificed after each sampling step for analysis (t= 0, 1, 2, 3, 4, 8 and 12 d). All incubations were conducted in triplicates.

3.2.2.2 Effect of operational parameters on fungal growth, pelletization and Se removal

Batch experiments were conducted to examine the effect of solution pH (2.5-7.0), concentration of glucose (0.5-10 g L^{-1}) and initial concentration of Se (2-10 mg Se L^{-1}) added as SeO_3^{2-} or SeO_4^{2-} on the removal of total soluble Se. To assess the effect of pH, the medium pH was adjusted by adding 0.1M KOH or 0.1 M HCl at the beginning of the experiment and not controlled afterwards. All experiments were conducted under the same conditions as above, varying one operational parameter at a time and keeping the others constant. The incubation period for these experiments was 4 d, which was observed to be the period at which most of the biomass was already produced.

3.2.3 Transmission electron microscopy (TEM) and electron-energy loss spectroscopy (EELS) analysis

Fixed (glutaraldehyde 2.5% in phosphate buffer saline, pH 7) fungal samples were placed on a carbon coated copper grid and stained with 0.5% uranyl acetate (Burghardt and Droleskey, 2005). Imaging of samples was performed on a Titan G2 80–300 kV transmission electron microscope (FEI Company, The Netherlands) equipped with a 4 k × 4 k CCD camera model US4000 and an energy filter model GIF Tridiem (Gatan, Inc.). The general morphology and size of the produced Se^0 particles were determined using TEM images. Particle size measurements were done using Image J software (version 1.47, National Institute of Health, USA) (Rasband, 1997-2014) based on TEM images. The average particle size was calculated from measuring particles in random fields of view (Teodoro et al., 2011). To obtain the map distribution of Se in the samples, the EELS signal from Se L edge (Se-L edge of 1436 eV) and M edge (Se-M edge of 57 eV) was acquired in energy-filtered TEM (EFTEM) mode. Each elemental map was created by using a 3-window method (Kim and Dong, 2011).

To determine the location of the Se^0 particles within the fungal biomass, a 3D reconstruction was done. The sample was imaged using a Titan CT operating at 300 kV equipped with a 4k CCD camera (Gatan, Pleasanton, CA, USA). Tilt series for tomographic reconstruction were acquired using the Xplore 3D tomography software (FEI Company). Using a tilt range of ± 68°, images were captured using a Saxton scheme at 2° intervals (Koning and Koster, 2013). The tomogram was generated using the Weighted-Back Projection algorithm as implemented in the IMOD software (Kremer et al., 1996). Segmentation and three-dimensional rendering of the tomographic images was conducted using Avizo (Visualization Science Group) image-processing software.

3.2.4 Analytical methods

After the incubation period, samples were centrifuged at 6000 rpm for 15 min. The supernatant was used for measuring the chemical oxygen demand (COD), glucose and the total soluble residual Se concentration in solution. The COD was determined according to the APHA 5220D standard procedure (APHA, 1995). The glucose concentration was analyzed with the dinitrosalicylic acid method using D-glucose as standard (Miller, 1959). For the measurement of the total Se concentration, the samples were further filtered (0.45 μm) and preserved with an acidified solution of 0.5% HNO_3 in ultrapure water (Milli-Q water, 18MΩ-cm). Se was analyzed by inductively coupled plasma spectrometry (ICP-MS), with H_2:He (7:93) as the reaction gas, and ^{78}Se and ^{80}Se used for quantification and verification, respectively. Lithium, gallium, scandium, rhodium and iridium were used as internal standards. Samples were injected in 1:1 ratio, and measured three times. The system was flushed entirely with an acidified solution (0.5% HNO_3 in Milli-Q water).

After centrifugation, the fungal biomass was collected and washed several times with ultrapure water (Milli-Q). The biomass concentration was determined gravimetrically as dry weight by filtering the biomass suspension through a pre-dried (24 h at 105 °C) and pre-weighted filter paper (Type GF/F, Whatman Inc., Florham Park, NJ) of 0.45 μm pore-size. To count the number of pellets, samples were taken from the Erlenmeyer flasks after the incubation period and transferred manually into Petri dishes. The average size of the pellets was estimated with a Vernier caliper.

3.3 Results

3.3.1 Fungal interaction with Se oxyanions

The presence of Se oxyanions in the culture medium caused different effects on the growth, substrate consumption and pellet morphology of *P. chrysosporium* after 12 days of incubation. The biomass yield was clearly reduced in incubations with SeO_3^{2-} in comparison to incubations with SeO_4^{2-} and the biotic control (Table 3.1). The addition of SeO_3^{2-} at 10 mg Se L^{-1} inhibited the fungal growth and resulted in an 80% decrease of the biomass concentration compared to the Se-free incubations (Figure 3.1A). In contrast, SeO_4^{2-} was less inhibitory than SeO_3^{2-} to the fungal growth. In the presence of SeO_4^{2-} at 10 mg Se L^{-1}, a 30% decrease of the biomass concentration compared to the biotic control was observed (Figure 3.1A).

Table 3.1 Effects of Se oxyanions on the fungal growth and morphology of *P. chrysosporium* over 12 d of incubation.

Incubation	Shape	Surface	Color	Dry weight ($g\ L^{-1}$)[a]	$Y_{x/s}$ ($g\ g^{-1}$)[a, b]
Control	Pellet	Hairy	Ivory white	1.51 ± 0.04	0.23±0.005
SeO_3^{2-}	Pellet	Smooth	Red-orange	0.27±0.05	0.11±0.04
SeO_4^{2-}	Pellet	Hairy and smooth	Ivory white	1.12 ± 0.04	0.21±0.01

[a] Values are means (n=3) with standard deviations

[b] Biomass yield. Calculated as $Yx_{/s} = \frac{\Delta x}{\Delta s}$

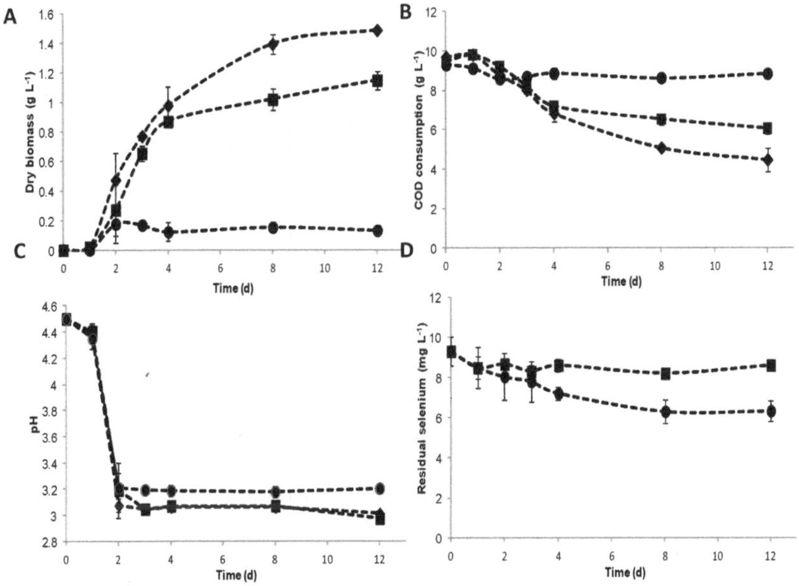

Figure 3.1 A) Growth of biomass in the biotic control (Se-free) or exposed to selenate and selenite (10 mg Se L^{-1}), B) COD consumption, C) pH profile, and D) Se removal. The symbols refer to: ◆ Control, ■ SeO_4^{2-}, ● SeO_3^{2-}.

Substrate consumption was also decreased in the presence of SeO_3^{2-}, only about 30% of the glucose was consumed (Figure 3.1B). In the presence of SeO_4^{2-} and when Se oxyanions were omitted (*i.e.,* biotic control) the glucose consumption was between 55-65% (Figure 3.1B). For all incubations, the pH was between 3.0-3.4 after the second day of incubation (Figure 3.1C). The removal of Se (Figure 3.1D) from the growth medium was higher for the SeO_3^{2-} incubation (40% removal) compared to the SeO_4^{2-} incubation (< 10% removal). The abiotic controls (no biomass) did not show any changes from the initial conditions over 12 days of incubation (data not shown), indicating that the observed changes in the Se removal were due to the metabolic activity of *P. chrysosporium*. Moreover, no change in the Se concentration was observed for controls with dead biomass after the incubation period, which suggests that the decrease of the Se concentration with live biomass was not attributed to adsorption (data not shown). Incubations with SeO_3^{2-} presented a garlic like-odor, suggesting the formation of volatile forms of Se (Gharieb et al., 1995).

The morphology of the fungal pellets differed between incubations (Table 3.1, Figure 3.2). The formation of fungal pellets was observed in all the incubations (except for the dead control) irrespective of the presence or absence of Se. However, when exposed to SeO_3^{2-}, pellets of *P. chrysosporium* were compact and smooth, with a characteristic red-orange color, which indicates the reduction of SeO_3^{2-} to Se^0 (Gharieb et al., 1995; Sarkar et al., 2011). The number of pellets per incubation (<18) was lower and the size of the pellets was in average smaller for the SeO_3^{2-} incubations (Figure 3.2) compared to the other incubations. The morphology of pellets with SeO_4^{2-} was irregular, showing both smooth and hairy pellets, with an ivory white color (Figure 3.2). In contrast, the pellets grown in the absence of Se were hairy, fluffy and with an ivory white color, as reported before for this fungus (Huang et al., 2010). The number of pellets (>25) was higher and their size was larger (Figure 3.2) for the biotic control (Se-free incubations). It is important to highlight that reproducibility of fungal pelletization in the presence of SeO_3^{2-} was achieved in all the experiments.

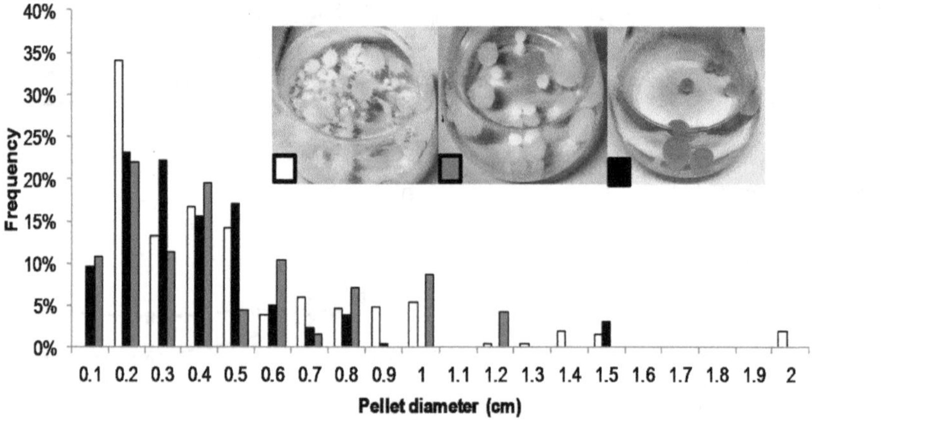

Figure 3.2 Appearance and size distribution of pellets of *P. chrysosporium* produced in the absence of Se (■ control), with 10 mg L^{-1} of selenate (□ SeO_4^{2-}) or selenite (■ SeO_3^{2-}) after 12 days of incubation.

The reduction of Se oxyanions to Se^0 was indicated by visualization and further confirmed by microscopic imaging. A red-orange coloration, suggesting the formation of Se^0, was only observed when the fungus was exposed to SeO_3^{2-}. Confirmation of the production of Se^0 particles by the fungal biomass was obtained with TEM-EELS analysis. From TEM imaging, Se^0 particles were not equally distributed in the hyphae (Figures 3.3A, B and C), suggesting that Se^0 formation is specifically localized. Image analysis showed different size diameters for the produced particles, in a range of 35-400 nm, with the majority (>60%) of the structures being *true* nanoparticles (<100 nm). A 3D reconstruction (Figure 3.4) using a series of TEM images confirmed that the Se^0 particles were located within the fungal biomass.

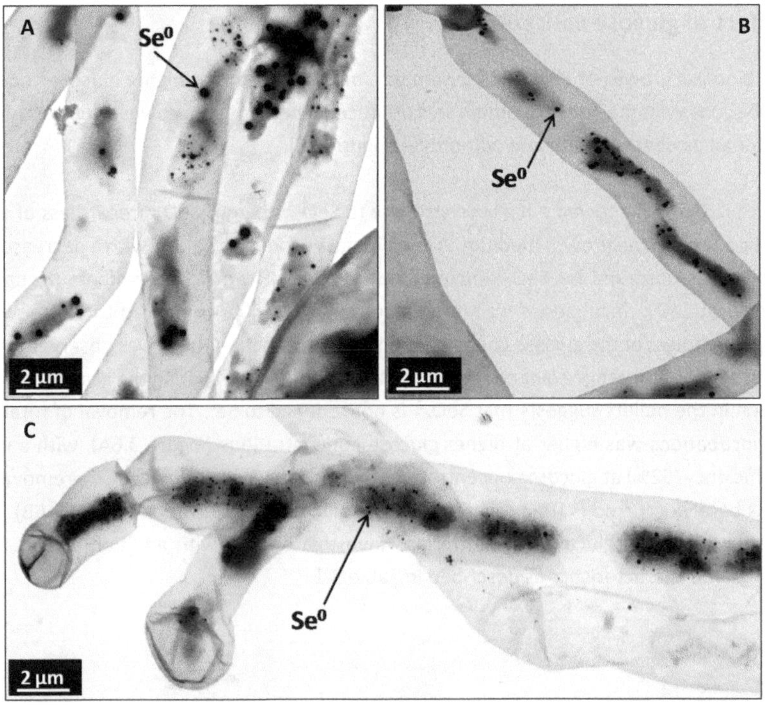

Figure 3.3 Transmission electron micrographs of nSe^0 produced in the biomass of *P. chrysosporium*. a) Distribution of Se^0 particles of different sizes within fungal biomass, b) and c) Localization of Se^0 particles within fungal biomass.

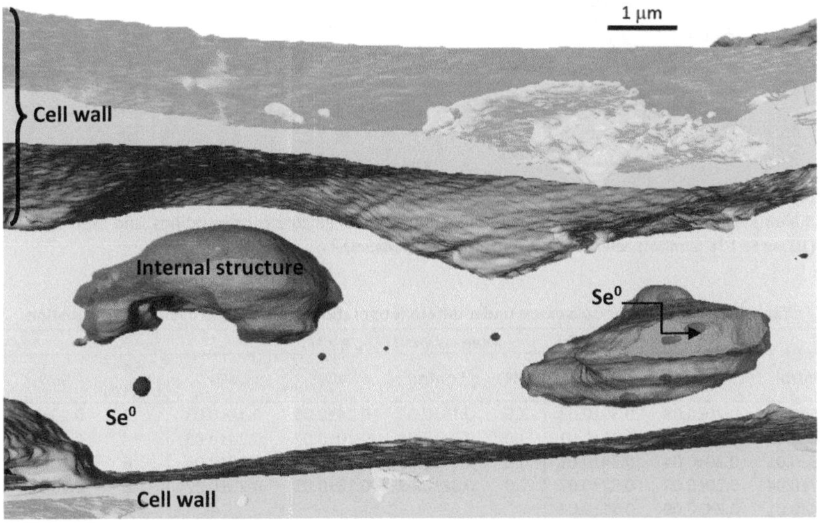

Figure 3.4 3D reconstruction model of Se^0 nanoparticles within fungal biomass.

3.3.2 Effect of glucose concentration

The biomass growth of *P. chrysosporium* was more abundant at higher glucose concentrations for all incubations. Minimal fungal biomass was produced when using the lowest glucose concentration (0.5 g L^{-1}) for all incubations with and without Se (Figure 3.5).

P. chrysosporium showed a higher sensitivity to SeO_3^{2-} than to SeO_4^{2-} regardless of the glucose concentration used in the growth medium. A low biomass production along with a decrease of glucose consumption was observed for SeO_3^{2-} incubations compared to other treatments (Figure 3.5). The biomass yield of SeO_3^{2-} incubations was in all cases less than 60% compared to Se-free incubations (Table 3.2). Regardless of the glucose concentration, SeO_3^{2-} reduction to Se^0 was observed; pellets were smooth and with a red-orange color characteristic to the synthesis of Se^0. In the case of SeO_4^{2-}, the lack of coloration in the pellets suggests that SeO_4^{2-} is not reduced to Se^0. The removal of total soluble Se for SeO_3^{2-} incubations was higher at higher glucose concentrations (Figure 3.6A), with a maximal Se removal efficiency (52%) at glucose concentrations exceeding 7.5 g L^{-1}. However, Se removal for SeO_4^{2-} incubations barely occurred (<10%) regardless of the glucose concentration (Figure 3.6B). The fungal pellet morphology at all glucose concentrations investigated and with addition of SeO_3^{2-} or SeO_4^{2-} showed the same characteristics as described in Table 3.1.

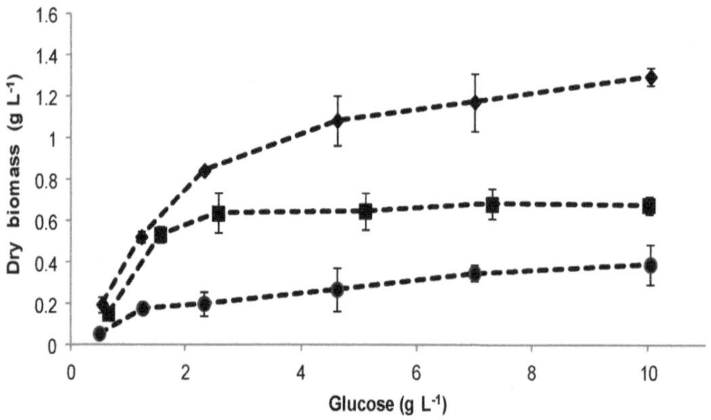

Figure 3.5 Biomass growth of *P. chrysosporium* as a function of initial glucose concentrations, and exposed to selenate or selenite (10 mg Se L^{-1}). Symbols refer to: ◆ Control, ■ SeO_4^{2-}, ● SeO_3^{2-}.

Table 3.2 Yield of *P. chrysosporium* under different operational parameters over 4 d of incubation.

Biomass yield ($Y_{x/s}$, g g^{-1})										
Glucose (g L^{-1})	Control	SeO_3^{2-}	SeO_4^{2-}	pH	Control	SeO_3^{2-}	SeO_4^{2-}	Se (mg L^{-1})	SeO_3^{2-}	SeO_4^{2-}
0.5	0.45±0.09	0.20±0.03	0.30±0.03	2.5	0.15±0.02	0.14±0.06	0.16±0.03	2	0.25±0.02	0.36±0.03
1.5	0.46±0.02	0.26±0.06	0.41±0.01	3.0	0.27±0.04	0.10±0.02	0.24±0.05	4	0.20±0.02	0.36±0.03
2.5	0.39±0.01	0.20±0.04	0.34±0.01	4.5	0.33±0.05	0.20±0.04	0.28±0.08	6	0.19±0.05	0.35±0.01
5.0	0.37±0.04	0.20±0.04	0.32±0.07	7.0	0.32±0.03	0.16±0.05	0.26±0.02	8	0.25±0.04	0.35±0.01
7.5	0.34±0.01	0.23±0.09	0.31±0.04					10	0.21±0.05	0.34±0.02
10	0.27±0.01	0.15±0.04	0.25±0.02							
Conditions for this experiment: pH 4.5, 10 mg Se L^{-1}				Conditions for this experiment: Glucose 10 g L^{-1}, 10 mg Se L^{-1}				Conditions for this experiment: Glucose 10 g L^{-1}, pH 4.5		

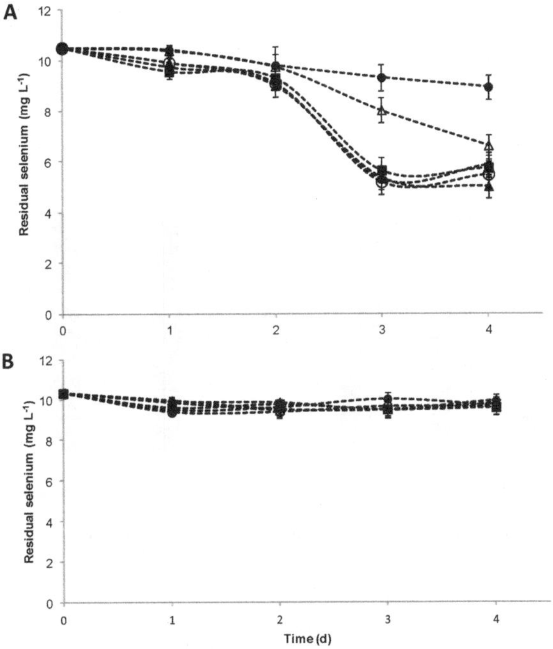

Figure 3.6 Removal of Se oxyanions (10 mg Se L^{-1}) at different glucose concentrations. A) Selenite and B) Selenate incubations. Symbols refer to: ● 0.5, △ 1.5, ■ 2.5, ◆ 5, ○ 7.5, ▲ 10 mg glucose L^{-1}.

3.3.3 Effect of pH

When the fungi were grown at different initial pH values, the maximal biomass growth occurred at pH 4.5 for all the incubations (Figure 3.7A). In general, the production of biomass was strongly inhibited under the most acidic conditions. At pH 2.5, there was about 70% less biomass produced for the SeO_4^{2-} and Se-free incubations, and about 90% less biomass for SeO_3^{2-} compared to the maximum biomass growth obtained at pH 4.5 (Figure 3.7A). At pH 3.0, a slight decrease in biomass concentration was found for SeO_4^{2-} and Se-free incubations compared to pH 4.5; whereas for SeO_3^{2-} fungal growth was similarly inhibited as when grown at pH 2.5. For all incubations, biomass yields obtained at pH 7.0 were similar to those obtained at pH 4.5 (Table 3.2).

The formation of Se^0 was not observed for any of the SeO_4^{2-} incubations at different pH conditions (Figure 3.7B). In contrast, the reduction of SeO_3^{2-} to Se^0 occurred over the whole pH range investigated (2.5-7.0). However, the removal of SeO_3^{2-} was pH dependent (Figure 3.7B): under the most acidic conditions (pH 2.5-3.0), the percentage of Se removal was only 10%, whereas at pH 4.5 and 7.0 the Se removal efficiency was about 45%.

The growth of the fungus under different pH conditions lead to morphological changes of the growing mycelium (Figure 3.8). Mycelia from Se-free media grown at low pH (2.5 and 3) exhibited an irregular short-strip shape as well as pellets; whereas regular mycelial pellets were observed at pH 4.5 and 7. The mycelia cultivated in the presence of SeO_3^{2-} showed less red-orange coloration at low pH (2.5 and 3) compared with those grown at pH 4.5 and 7. During incubations at pH 2.5, some pellets did even not show any sign of red-orange coloration, which indicated that the reduction of SeO_3^{2-} to Se^0

was limited. Incubations with SeO_4^{2-} only showed a sign of perturbation at pH 7, forming pellets and clumps of rough surface (Figure 3.8).

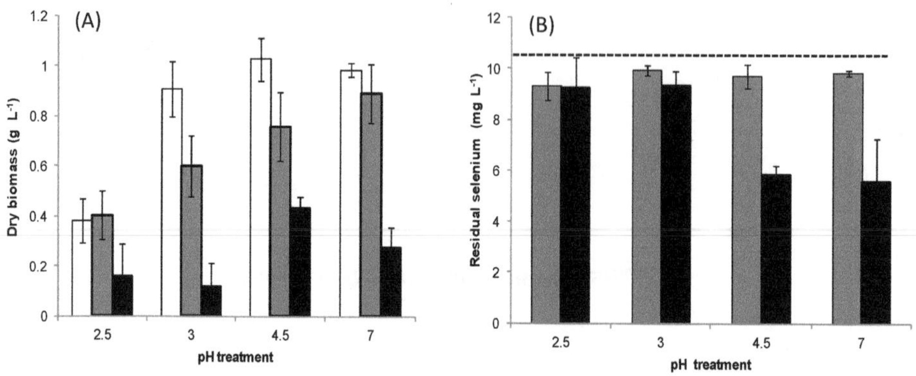

Figure 3.7 Response of *P. chrysosporium* to Se oxyanions under different pH conditions. A) Production of fungal biomass (dry weight), B) Se removal efficiency. Symbols refer to: ◻ Control, ◼ SeO_4^{2-}, ■ SeO_3^{2-}, ---- Initial Se concentration.

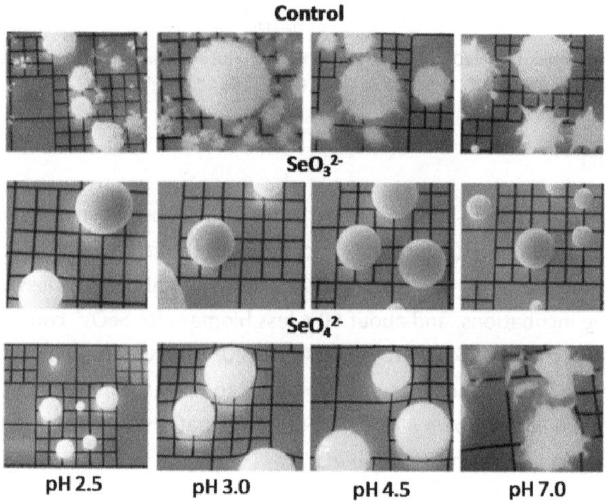

Figure 3.8 Effect of growth medium pH on the morphology of *P. chrysosporium* pellets.

3.3.4 Effects of Se concentration

Fungal growth was inhibited with increasing Se concentrations, for both treatments with SeO_4^{2-} and SeO_3^{2-}. Compared to the Se-free incubations, the biomass concentration was reduced by 44-80% at different doses of SeO_3^{2-} (Figure 3.9A), whereas the biomass concentration was reduced by 11-24% at increasing SeO_4^{2-} concentrations. The consumption of glucose was the highest for the incubations with lower Se concentrations (data not shown). The pH profile was similar for all the incubations, decreasing from initial pH 4.5 to 3.2 ±0.1.

The removal of Se oxyanions was not concentration dependent; it remained proportionally the same under all Se oxyanion concentrations tested (Figure 3.9B). The removal efficiency of total soluble Se was about 10% and 40% for incubations with SeO_4^{2-} and SeO_3^{2-}, respectively. No major morphological changes, in comparison to the already described characteristics (Table 3.1), were observed for the mycelia grown at different concentrations of SeO_3^{2-} or SeO_4^{2-}. Only, at low concentrations of SeO_3^{2-} (2-4 mg L^{-1}), a less intense red-orange coloration in the pellets was observed (data not shown).

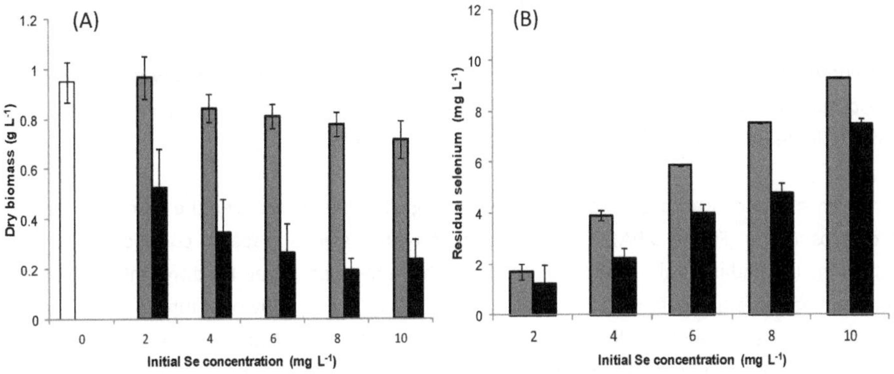

Figure 3.9 Response of *P. chrysosporium* to different concentrations of Se oxyanions. A) Production of fungal biomass (dry weight), B) Se removal efficiency. Symbols refer to: Control, SeO_4^{2-}, SeO_3^{2-}.

3.4 Discussion

3.4.1 Inhibition of fungal growth induced by Se oxyanions

The sensitivity of *P. chrysosporium* to Se in terms of dry biomass production was the highest when Se was added in the form of SeO_3^{2-}. A clear inhibitory effect of the fungal growth was observed even at SeO_3^{2-} concentrations as low as 2 mg Se L^{-1}. The inhibitory effect of SeO_3^{2-} to the fungal growth of filamentous, polymorphic and unicellular fungi has been demonstrated before (Gharieb et al., 1995). Addition of SeO_4^{2-}, which is the least toxic Se oxyanion on an acute basis (Brix et al., 2001; Canton, 1999), caused a less inhibitory effect on the fungal growth. In agreement with the results of the present study, previous investigations showed that the addition of SeO_4^{2-} had less impact on the inhibition of the growth of other white-rot fungi, such as *Bjerkandera adusta* (Catal et al., 2008), in comparison to SeO_3^{2-}. However, fungal tolerance to different Se oxyanions is species dependent (Gharieb et al., 1995). The growth of *Fusarium* sp., a Se-reducing fungus, was more inhibited in the presence of SeO_4^{2-} than in the presence of SeO_3^{2-} (Gharieb et al., 1995). In the case of *Pleurotus ostreatus*, 2.5 mg L^{-1} of SeO_3^{2-} stimulated the fungal growth of mycelium, whereas 5 mg L^{-1} was inhibitory (Serafin-Muñoz et al., 2006). The tolerance to Se is attributed to the incapacity of fungi to distinguish between Se and S. It has been suggested that Se toxicity is due to the incorporation of Se into sulfur-containing aminoacids and proteins instead of sulfur, altering their structure and disrupting the enzyme activity (Golubev and Golubev, 2002; Lauchli, 1993).

The composition of the growth medium can also influence the tolerance to Se, and therefore the fungal growth, particularly the presence of a certain carbon source (*e.g.,* glucose), sulfate, sulfur-containing amino acids or glutamine (Golubev and Golubev, 2002; Serafin-Muñoz et al., 2006). Glucose is the most readily utilizable substrate by fungi; it is of crucial importance for the optimal growth of fungi and it has been suggested to have an active role in SeO_3^{2-} uptake (Gharieb and Gadd, 2004). As it has been demonstrated in previous studies (Kim et al., 2003), the variation of the initial glucose concentration in the medium has a noticeable effect on the growth of fungi and biomass yield (Table 3.2). Indeed, above 5 g L^{-1}, increasing the glucose concentration did not have a prominent effect on the fungal growth when SeO_3^{2-} was added to the medium (Figure 3.5). The use of high concentrations of glucose (10 g L^{-1}) is common for the production of fungal biomass, however, such concentrations are not commonly found in wastewaters. A clear reduction of the glucose uptake was observed when *P. chrysosporium* was fed with Se oxyanions (Figure 3.1, Table 3.1), particularly with SeO_3^{2-}, which indicates a decrease of the carbohydrate metabolism in the cells.

Increased concentrations of Se oxyanions in the growth media reduced the biomass production of *P. chrysosporium* (Figure 3.9A). This is consistent with previous research conducted with other fungal strains (Ramadan et al., 1988). Similar effects have been observed for different fungal species exposed to heavy metals (Graz et al., 2011; Kim et al., 2003). The sensitivity of fungi to toxic elements (*i.e.,* heavy metals) is species dependent, which is mainly attributed to different detoxification mechanisms utilized by various fungi (Kim et al., 2003). Fungal growth was also influenced by pH. In comparison to other microorganisms, it is well known that a low pH is favorable for the growth of fungi. This study showed that the maximal fungal growth of *P. chrysosporium* was achieved at pH 4.5 for all incubations with and without Se (Figure 3.7A).

3.4.2 Morphological effects induced by Se oxyanions

This study shows that the presence of SeO_3^{2-} in the growth medium of *P. chrysosporium* induces growth stress to this organism (Figure 3.1A) which resulted in compact and smooth pellets of smaller diameter than pellets formed in the absence of SeO_3^{2-} (Figure 3.2). Certain species of filamentous fungi are able to grow in the form of pellets under submerged conditions in the liquid medium. Pellets are usually formed as a result of the interaction between hyphae, solid particles and spores (Prosser, 1995). Even though the exact mechanism for fungal pelletization remains unclear, it is well known that pelletization is driven by cultivation conditions (*e.g.,* pH, temperature, agitation rate), composition of the growth medium (*e.g.,* carbon source, nutrients, additives, carriers) and inoculum (*e.g.,* age and size of inoculum) (Metz and Kossen, 1977; Papagianni, 2004). Self-immobilization of fungi as pellets seems to occur mainly under stressful conditions, such as limitation of nutrients or oxygen. The presence of toxic compounds also influences the formation and characteristics of the pellets. Saraswathy and Hallberg (2005) showed that the presence of pyrene in the growth medium resulted in the formation of pellets by two *Penicillium ochrochloron* strains, and the size and texture of the pellets formed varied for each individual strain.

3.4.3 Removal of Se oxyanions by *P. chrysosporium*

The present study demonstrates that *P. chrysosporium* possesses Se-reducing capabilities. The main fungal species that are reported with Se-reducing capabilities include *Alternaria alternata* (Sarkar et al., 2011), *Aspergillus* spp. (Moss et al., 1987; Gharieb et al., 1995), *Fusarium* sp. (Ramadan et al., 1988; Gharieb et al., 1995), *Mortierella* spp. (Zieve et al., 1985), *Penicillium* spp. (Brady et al., 1996) and *Lentinula edodes* (Vetchinkina et al., 2013). Se removal (40-50%) by *P. chrysosporium* mainly occurred when supplied SeO_3^{2-}. This can be attributed to SeO_3^{2-} reduction to Se^0 as a detoxification mechanism (Gharieb et al., 1995). Se tolerance and detoxification by fungi have been mainly ascribed to: biomethylation of Se compounds, which involves the reduction of inorganic Se forms to less toxic and volatile derivatives, such as dimethylselenide (Gadd, 1993), or the reduction of Se oxyanions to Se^0, producing intra or extracellular red-orange deposits of Se^0 (Gharieb et al., 1995). The detection of a garlic-like odor in SeO_3^{2-} incubations suggests that volatile Se compounds are formed during the reduction to Se^0 (Gharieb et al., 1995). Further research is required to quantify the volatile fraction and the speciation of Se in the gas phase.

The detailed mechanism of selenate and selenite reduction by *P. chrysosporium* is not known. It has been reported that glutathione, a common intracellular reduced thiol in various organisms, including fungi, is involved in detoxification processes. A recent study showed that *P. chrysosporium* accumulated high levels of glutathione when exposed to high concentrations of heavy metals (Xu et al., 2014). It has been proposed that in some bacteria, selenite reacts with glutathione producing selenodiglutathione which is reduced by glutathione reductases to form a seleniumpersulfide compound. This compound then dismutates into Se^0 and reduced glutathione (Debieux et al., 2011). It is possible that *P. chrysosporium* uses a similar glutathione-dependent mechanism for the reduction of selenite. However, further studies involving proteomics would be required to test this hypothesis and to determine the reductases involved in the formation of the Se^0 nanoparticles.

The removal of SeO_3^{2-} by *P. chrysosporium* was influenced by the glucose concentration. Glucose concentrations higher than 2.5 g L^{-1} showed maximal removal of this Se oxyanion. A previous study demonstrated the stimulatory effect of glucose addition on SeO_3^{2-} accumulation by the yeast *Saccharomyces cerevisiae*, suggesting a predominant role of the metabolic activity to transport SeO_3^{2-} by the cells (Gharieb and Gadd, 2004). The positive influence of glucose on the uptake of SeO_3^{2-} has also been observed in bacteria, *i.e., Salmonella typhimurium* (Brown and Shrift, 1980).

The pH also played an important role on the fungal removal of Se as SeO_3^{2-} (Figure 3.7B). Similar maximum removal efficiencies of SeO_3^{2-} were obtained between pH 4.5 and 7.0, suggesting that the treatment of selenite-containing wastewater using fungi can be applied to both acidic and neutral waste streams. This feature represents a notable advantage over the reduction of SeO_3^{2-} to Se^0 by bacteria, since the bacterial process occurs at near to neutral up to alkaline pH (Lortie et al., 1992; Mishra et al., 2011).

Se removal was less than 10% for SeO_4^{2-} incubations, regardless of the glucose concentration (Figure 3.5), pH (Figure 3.7A) or initial Se concentration (Figure 3.9A). Therefore, the use of fungal pellets of *P. chrysosporium* to remove SeO_4^{2-} from wastewater is not recommended. The lack of red-orange coloration in the fungal biomass suggested that *P. chrysosporium* was not able to reduce SeO_4^{2-} to Se^0, as it has been reported also for other fungi (Vetchinkina et al., 2013). Even though the Se speciation was not determined in this study, the results suggest that the capacity of this fungus to reduce SeO_4^{2-} to SeO_3^{2-} is limited, as the overall soluble total Se removal was less than 10%. If this had been transformed into SeO_3^{2-}, then SeO_3^{2-} would have been further reduced to Se nanoparticles (the fungus readily reduced SeO_3^{2-} to selenium nanoparticles). However Se nanoparticles were not detected in these incubations. Moreover, if the selenate had been transformed into SeO_3^{2-}, then the fungal growth would have been inhibited similarly as for the incubations with SeO_3^{2-} (Figure 3.1A).

3.4.4 Production of Se^0 by *P. chrysosporium*

The characteristic orange-red coloration that results from Se^0 synthesis (Figure 3.2) was only observed in the biomass of the SeO_3^{2-} incubations and not in the medium, indicating immobilization of Se^0 in the fungal material. Deposition and entrapment of Se^0 within fungal biomass might be advantageous for technical applications, considering that there will be no loss or washout of any material. Formation of *true* nanoparticles of Se^0 (particles of size <100 nm) was also observed (Figure 3.3). The synthesis of Se^0 from SeO_3^{2-} by different bacterial strains such as *Bacillus megaterium*, *Sulfurospirillum barnessi* and *Selenihalanaerobacter shriftii* has also been reported (Oremland et al., 2004; Mishra et al., 2011). However, most of the particles produced by bacteria are much larger in size (200-1000 nm) compared to those produced by *P. chrysosporium* (35-400 nm).

TEM cell characterization and the corresponding 3D reconstruction (Figure 3.4) confirmed that the majority of the Se^0 particles were located inside the fungal cells and that some were localized within the fungal cell wall, suggesting their intracellular formation. Mukherjee et al. (2001) suggested that intracellular production of nanoparticles is driven by the electrostatic interaction between the metal ions in solution and the enzymes in the fungal cell wall, binding on the fungal cell surface, where the metal ions are reduced, leading to the synthesis of nanoparticles that accumulate within the mycelia. Even though the majority of the nanoparticles synthesized by fungi have been observed to be formed extracellularly (Verma et al., 2010; Syed and Ahmad, 2012), a few species such as *Verticillium* sp. (Mukherjee et al., 2001; Sastry et al., 2003), *Thrichothecium* sp. (Ahmad et al., 2005) and *Aspergillus flavus* (Vigneshwaran et al., 2007; Rajakumar et al., 2012) synthesize nanoparticles intracellularly. From the TEM images (Figure 3.3), it seems that the majority of nanoparticles are compartmentalized in the fungal cell, in intracellular structures. Further analyses need to be performed in order to determine specifically in which organelles or structures the synthesis of Se^0 is taking place. Some fungal species (*Aspergillus funiculosus* and *Fusarium* sp.) have shown the ability to compartmentalize Se^0 in their vacuoles (Gharieb 1993; Gharieb et al., 1995), which have been suggested to regulate the uptake, detoxification and tolerance to SeO_3^{2-} in the yeast *S. cerevisiae*.

3.4.5 Potential applications

Removal of Se oxyanions at low pH is promising for the treatment of acidic effluents polluted with Se, particularly as SeO_3^{2-}. *Phanerochaete chrysosporium* possesses the ability to intracellularly produce nSe^0, which can be separated from the treated effluent in an immobilized form suitable for use in commercial applications. There is an increased interest in the use of nSe^0, including applications such as antifungal and anti-cancer agents (van Cutsem et al., 1990; Ahmad et al., 2010) as well as an effective agent to prevent and treat *Staphylococus aureus* infections (Tran and Webster, 2011). Besides, biogenic nSe^0 have been used for the production of high sensitivity sensors (Wang et al., 2010) and as potential affinity sorbents for contaminants such as mercury (Fellowes et al., 2011) and zinc (Jain et al., 2014). Moreover, the influence of Se (SeO_3^{2-}) on the growth and pelletization of *P. chrysosporium* could be of potential application to control biomass growth in fungal bioreactors.

Moreover, SeO_3^{2-} adds to the repertoire of factors that influence pellet formation and fungal morphology.

3.5 References

Ahmad A., Senapati S., Khan M.I., Kumar R., Sastry M. (2005) Extra/intracellular, biosynthesis of gold nanoparticles by an alkalotolerant fungus. *Trichothecium* sp. J Biomed Nanotechnol 1:47–53.

Ahmad R.S., Ali F., Gazal M., Parisa J.F., Sassan R., Seyed M.R. (2010) Antifungal activity of biogenic selenium nanoparticles. World Appl Sci J 10(8):912–922.

APHA (1995) Standard Methods for Water and Wastewater Examination. 19th ed. American Public Health Association, Washington, DC, USA.

Barkes L., Fleming R.W. (1974) Production of dimethylselenide gas from inorganic selenium by eleven soil fungi. Bull Environ Contam Toxicol 12:308–311.

Beheshti N., Soflaei S., Shakibaie M., Yazdi M.H., Ghaffarifar F., Dalimi A., Shahverdi A.R. (2013) Efficacy of biogenic selenium nanoparticles against *Leishmania major*: in vitro and in vivo studies. J Trace Elem Med Biol 27(3):203–207.

Bleiman N., Mishael Y.G. (2010) Selenium removal from drinking water by adsorption to chitosan-clay composites and oxides: Batch and columns tests. J Hazard Mat 183(1-3):590–595.

Brady J.M., Tobin J.M., Gadd G.M. (1996) Volatilization of selenite in aqueous medium by a *Penicillium* species. Mycol Res 100:955–961.

Brix K.V., Adams W.J., Reash R.J., Carlton R.G., McIntyre D.O. (2001) Acute toxicity of sodium selenate to two daphnids and three amphipods. Environ Toxicol 16(2):142–150.

Brown T.A., Shrift A. (1980) Assimilation of selenate and selenite by *Salmonella typhimurium*. Can J Microbiol 26:671–675.

Cameron M.D., Timofeevski S., Aust S.D. (2000) Enzymology of *Phanerochaete chrysosporium* with respect to the degradation of recalcitrant compounds and xenobiotics. Appl Microbiol Biotechnol 54:751–758.

Canton S.P. (1999) Acute aquatic life criteria for selenium. Environ Toxicol Chem 18:1425–1432.

Castro-Longoria E., Vilchis-Nestor A., Avalos-Borja M. (2011) Biosynthesis of silver, gold and bimetallic nanoparticles using the filamentous fungus *Neurospora crassa*. Colloids Surf B Biointerfaces 83:42–48.

Catal T., Liu H., Bermek H. (2008) Selenium induces Manganese-dependent peroxidase production by the white-rot fungus *Bjerkandera adusta* (Willdenow) P. Karsten. Biol Trace Elem Res 123:211–217.

Chan Y.S., Don M.M. (2013) Biosynthesis and structural characterization of Ag nanoparticles from white-rot fungi. Mat Sci Eng C 33:282–288.

Debieux C.M., Dridge E.J., Mueller C.M., Splatt P., Paszkiewicz K., Knight I., Florance H., Love J., Titball R.W., Lewis R.J., Richardson D.J., Butler C.S. (2011) A bacterial process for selenium nanosphere assembly. Proc Natl Acad Sci USA 108:13480–13485.

Fleming R.W., Alexander M. (1972) Dimethylselenide and dimethyltelluride formation by a strain of *Penicillium*. Appl Microbiol 24:424–429.

Fordyce F.M. (2013) Selenium deficiency and toxicity in the environment. In: Selinus O (ed) Essentials of Medical Geology, revised edn. Springer, Netherlands, pp. 375–416.

Frankenberger W.T., Amrhein C., Fan T.W.M., Flaschi D., Glater J., Kartinen E., Kovac K., Lee E., Ohlendorf H.M., Owens L., Terry N., Toto A. (2004) Advanced treatment technologies in the remediation of seleniferous drainage waters and sediments. Irrig Dran Syst. 18:19–41.

Gadd G.M. (1993) Microbial formation and transformation of organometallic and organometalloid compounds. FEMS Microbiol Rev 11:297–316.

Gharieb M.M. (1993) Selenium toxicity, accumulation and metabolism by fungi and influence of the fungicide dithane. PhD thesis. University of Dundee.

Gharieb M.M., Wilkinson S.C., Gadd G.M. (1995) Reduction of selenium oxyanions by unicellular, polymorphic and filamentous fungi: cellular location of reduced selenium and implication for tolerance. J Ind Microbiol 14(3-4):300–311.

Gharieb M.M., Gadd G.M. (1998) Evidence for the involvement of vacuolar activity in metalloid tolerance: vacuolar-lacking and defective mutants of *Saccharomyces cerevisiae* display higher sensitivity to chromate, tellurite and selenite. Biometals 11:101–106.

Gharieb M.M., Gadd G.M. (2004) The kinetics of ^{75}[Se]-selenite uptake by *Saccharomyces cerevisiae* and the vacuolization response to high concentrations. Mycol Res 108(12):1415–1422.

Geoffroy N., Demopoulos G.P. (2011) The elimination of selenium (IV) from aqueous solution by precipitation with sodium sulfide. J Hazard Mater 185(1):148–154.

Golubev V.I., Golubev N.V. (2002) Selenium tolerance of yeasts. Microbiol 71(4):455-459.

Graz M., Pawlikowska-Pawlega B., Jarosz-Wilkolazka A. (2011) Growth inhibition and intracellular distribution of Pb ions by the white-rot fungus *Abortiporus biennis*. Int Biodeter Biodegr 65:124–129.

Fellowes J.W., Pattrick R.A., Green D.I., Dent A., Lloyd J.R., Pearce C.I. (2011) Use of biogenic and abiotic elemental selenium nanospheres to sequester elemental mercury released from mercury contaminated museum specimens. J Hazard Mater 189(3):660–669.

Hamilton S.J. (2004) Review of selenium toxicity in the aquatic food chain. Sci Total Environ 326(1-3):1–31.

Huang D.L., Zeng G.M., Feng C.L., Hu S., Zhao M.H., Lai C., Zhang Y., Jiang X.Y., Liu H.L. (2010) Mycelial growth and solid-state fermentation of lignocellulosic waste by white-rot fungus *Phanerochaete chrysosporium* under lead stress. Chemosphere 81:1091–1097.

Jain R., Jordan N., Schild, van Hullebusch E.D., Weiss S., Franzen C., Farges F., Hübner R,. Lens P.N.L. (2014) Adsorption of zinc by biogenic elemental selenium nanoparticles. Chem Eng J 260:855–863.

Kim J., Dong H. (2011) Application of electron energy-loss spectroscopy (EELS) and energy-filtered transmission electron microscopy (EFTEM) to the study of mineral transformation associated with microbial Fe-reduction of magnetite. Clays Clay Miner 59:176–188.

Kim C.G., Power S.A., Bell J.N.B. (2003) Effects of cadmium on growth and glucose utilization of ectomycorrhizal fungi in vitro. Mycorrhiza 13:223–226.

Koning R., Koster A. (2013) Cellular nanoimaging by cryo electron tomography. In: Sousa A.A., Kruhlak M.J. (eds), Nanoimaging: Methods and Protocols 950. Humana Press, Totowa, NJ. pp. 227–251.

Prosser J.I. (1995) Kinetics of filamentous growth and branching. In: Gow NAR, Gadd GM (eds), The growing fungus. Springer, The Netherlands. pp. 301–318.

Kremer J.R., Mastronarde DN, McIntosh JR (1996). Computer Visualization of Three-Dimensional Image Data Using IMOD. J Struct Biol 116:71–76.

Lauchli A. (1993) Selenium in Plants: uptake, functions and environmental toxicity. Bot Acta 106:455–468.

Lee H., Jang Y., Choi Y.S., Kim M.J., Lee J., Lee H., Hong J.H., Lee Y.M., Kim G.H., Kim J.J. (2014) Biotechnological procedures to select white rot fungi for the degradation of PAHs. J Microbiol Methods 97:56–62.

Lenz M., Lens P.N.L. (2009) The essential toxin: The changing perception of selenium in environmental sciences. Sci Total Environ 407:3620–3633.

Lortie L., Gould W.D., Rajan S., McCready R.G.L., Cheng K.J. (1992) Reduction of selenate and selenite to elemental selenium by a *Pseudomonas stutzeri* isolate. Appl Environ Microbiol 58:4042–4044.

Metz B., Kossen N.W.F. (1977) The growth of molds in the form of pellets – a literature review. Biotechnol Bioeng 19:781–99.

Mavrov V., Stamenov S., Todorova E., Chmiel H., Erwe T. (2006) New hybrid electrocoagulation membrane process for removing selenium from industrial wastewater. Desalination. 201:290–296.

Miller G.L. (1959) Use of dinitrosalicylic acid reagent for determination of reducing sugar. Anal Chem 31(3):426–428.

Mittal A.K., Chisti Y., Banerjee U.C. (2013) Synthesis of metallic nanoparticles using plant extracts. Biotechnol Adv 31(2):346–356.

Mishra R.R., Prajapati S., Das J., Dangar T.K., Das N., Thatoi H. (2011) Reduction of selenite to red elemental selenium by moderately halotolerant *Bacillus megaterium* strains isolated from *Bhitarkanika mangrove* soil and characterization of reduced product. Chemosphere 84:1231–1237.

Moldes D., Rodríguez S., Cameselle C., Sanromán M.A. (2003) Study of the degradation of dyes by MnP of *Phanerochaete chrysosporium* produced in a fixed-bed bioreactor. Chemosphere 51:295–303.

Moss M.O., Badii F., Gibbs G. (1987) Reduction of biselenite to elemental selenium by *Aspergillus parasiticus*. Trans Br Mycol Soc 89(4):578–580.

Mukherjee P., Ahmad A., Mandal D., Senapati S., Sainkar S.R., Khan M.I., Ramani R., Parischa R., Ajayakumar P.V., Alam M., Sastry M., Kumar R. (2001) Bioreduction of $AuCl_4^-$ ions by the fungus *Verticillium sp.* and surface trapping of the gold nanoparticles formed. Angew Chem Int Ed Engl 40:3585–3588.

Nguyen V.N.H., Amal R., Beydoun D. (2005) Photocatalytic reduction of selenium ions using different TiO_2 photocatalysts. Chem Eng Sci 60(21):5759–5769.

Oremland R.S., Herbel M.J., Blum J.S., Langley S., Beveridge T.J., Ajayan P.M., Sutto T., Ellis A.V. (2004) Structural and spectral features of selenium nanospheres produced by Se-respiring bacteria. Appl Environ Microbiol 70:52–60.

Papagianni M. (2004) Fungal morphology and metabolite production in submerged mycelial processes. Biotechnol Adv 22(3):189–259.

Pazouki M., Panda T. (2000) Understanding the morphology of fungi. Bioprocess Eng 22:127-143.

Prasad K.S., Patel H., Patel T., Patel K., Selvaraj K. (2013) Biosynthesis of Se nanoparticles and its effect on UV-induced DNA damage. Colloids Surf B Biointerfaces 103:261–266.

Prosser J.I. (1995) Kinetics of filamentous growth and branching. In: Gow NAR, Gadd GM (eds), The growing fungus. Springer, The Netherlands, pp. 301–318.

Rajakumar G., Rahuman A.A., Roopan S.M., Khanna V.G., Elango G., Kamaraj C., Zahir A.A., Velayutham K. (2012) Fungus-mediated biosynthesis and characterization of TiO_2 nanoparticles and their activity against pathogenic bacteria. Spectrochimica Acta Part A 91:23–29.

Ramadan S.E., Razak A.A., Youssef Y.A., Sedky N.M. (1988) Selenium metabolism in a strain of *Fusarium*. Biol Trace Elem Res 18:161–170.

Rasband W.S. (1997-2014) ImageJ. U.S. National Institute of Health, Bethesda, Maryland, USA. http://imagej.nih.gov/ij/. Accessed 14 May 2014.

Rayman M.P. (2012) Selenium and human health. Lancet 379:1256–1268.

Sanghi R., Verma P., Puri S. (2011) Enzymatic formation of gold nanoparticles using *Phanerochaete chrysosporium*. Adv Chem Eng Sci 1(3):154–162.

Saraswathy A., Hallberg R. (2005) Mycelial pellet formation by *Penicillium ochrochloron* species due to exposure to pyrene. Microbiol Res 160(4):375–383.

Sarkar J., Dey P., Saha S., Acharya K. (2011) Mycosynthesis of selenium nanoparticles. Micro Nano Lett 6(8):599–602.

Sastry M., Ahmad A., Islam N., Kumar R. (2003) Biosynthesis of metal nanoparticles using fungi and actinomycete. Current Sci 85(2):162–170.

Serafin-Muñoz A.H., Kubachka K., Wrobel K., Gutierrez-Corona J.F., Yathavakilla S.K.V., Caruso J.A., Wrobel K. (2006) Se-enriched mycelia of *Pleurotus ostreatus*: distribution of selenium in cell walls and cell membranes/cytosol. Agric Food Chem. 54:3440–3444.

Soda S., Ike M. (2011) Characterization of *Pseudomonas stutzeri* NT-I capable of removing soluble selenium from the aqueous phase under aerobic conditions. J Biosci and Bioeng 112(3):259–264.

Syed A., Ahmad A. (2012) Extracellular biosynthesis of platinum nanoparticles using the fungus *Fusarium oxysporum*. Colloids Surf B Biointerfaces 97:27–31.

Teodoro J.S., Simõesa A.M., Duarte F.V., Rolo A.P., Murdoch R.C., Hussain S.M., Palmeira C.M. (2011) Assessment of the toxicity of silver nanoparticles *in vitro*: A mitocohondrial perspective. Toxicol in Vitro 25(3):664–670.

Tien M., Kirk T.K. (1988) Lignin peroxidase of *Phanerochaete chrysosporium*. In: Wood WA, Kellogg ST (eds), Methods in Enzymology - Biomass, Part b, lignin, pectin and chitin. San Diego CA, Academic Press. Vol. 161, pp. 238–249.

Tran P.A., Webster T.J. (2011) Selenium nanoparticles inhibit *Staphylococcus aureus* growth. Int J Nanomedicine 6:1553–1558.

Tweedie J.W., Segel I.H. (1970) Specificity of transport processes for sulfur, selenium, and molybdenum anions by filamentous fungi. Biochim Biophys Acta 196(1):95–106.

Thongchul N., Yang S.T. (2003) Controlling filamentous fungal morphology by immobilization on a rotating fibrous matrix to enhance oxygen transfer and L(+)-lactic acid production by *Rhizopus oryzae*. In: Saha B.C. (ed). ACS Symposium Series 862, Fermentation Process Development. New York, Oxford University Press, pp. 36–51.

USHHS. United States Department for Health and Human Services. Toxicological Profile for Selenium; 2003. Available at: http://www.atsdr.cdc.gov/toxprofiles/tp92.pdf Accessed 05 March 2014.

van Cutsem J., Van Gerven F., Fransen J., Schrooten P., Janssen P.A. (1990) The in vitro antifungal activity of ketoconazole, zinc pyrithione, and selenium sulfide against *Pityrosporum* and their efficacy as a shampoo in the treatment of experimental pityrosporosis in guinea pigs. J Am Acad Dermatol 22(6 Pt 1):993–998.

Verma V.C., Kharwar R.N., Gange A.C. (2010) Biosynthesis of antimicrobial silver nanoparticles by the endophytic fungus *Aspergillus clavatus*. Nanomedicine 5:33–40.

Vetchinkina E., Loshchinina E., Kursky V., Nikitina V. (2013) Reduction of organic and inorganic selenium compounds by the edible medicinal basidiomycete *Lentinula edodes* and the accumulation of elemental selenium nanoparticles in its mycelium. J Micriobiol 51(6):829–835.

Vigneshwaran N., Kathe A.A., Varadarajan P.V., Nachane R.P., Balasubramanya R.H. (2006) Biomimetics of silver nanoparticles by white rot fungus, *Phanerochaete chrysosporium.* Colloids Surf B Biointerfaces 53:55–59.

Vigneshwaran N., Ashtaputre N.M., Varadarajan P.V., Nachane R.P., Paralikar K.M., Balasubramanya R.H. (2007) Biological synthesis of silver nanoparticles using the fungus *Aspergillus flavus.* Mater Lett 61:1413–1418.

Wang C., Sun H., Li J., Li Y., Zhang Q. (2009) Enzyme activities during degradation of polycyclic aromatic hydrocarbons by white rot fungus *Phanerochaete chrysosporium* in soils. Chemosphere 77:733–738.

Wang T., Yang L., Zhang B., Liu J. (2010) Extracellular biosynthesis and transformation of selenium nanoparticles and application in H_2O_2 biosensor. Colloids Surf B Biointerfaces 80(1):94–102.

Xu P., Liu L., Zeng G., Huang D., Lai C., Zhao M., Huang C., Li N., Wei Z., Wu H., Zhang C., Lai M., He Y. (2014) Heavy metal-induced glutathione accumulation and its role in heavy metal detoxification in *Phanerochaete chrysosporium.* Appl Microbiol Biotechnol 98:6409–6418.

Zelmanov G., Semiat R. (2013) Selenium removal from water and its recovery using iron (Fe^{3+}) oxide/hydroxide-based nanoparticles sol (NanoFe) as an adsorbent. Separ Purif Technol 103:167–172.

Zhen S., Su J., Wang L., Yao R., Wang D., Deng Y., Wang R., Wang G., Rensing C. (2014) Selenite reduction by the obligate aerobic bacterium *Comamonas testosteroni* S44 isolated from a metal-contaminated soil. BMC Microbiology 14(1):204.

Zhang J., Wang H., Yan X., Zhang L. (2005) Comparison of short-term *toxicity* between Nano-Se and selenite in mice. Life Sci 76:1099–1109.

Zhang J., Wang X., Xu T. (2008) Elemental selenium at nano size (Nano-Se) as a potential chemopreventive agent with reduced risk of selenium toxicity: comparison with se-methylselenocysteine in mice. Toxicol Sci 101(1):22–31.

Zhang W., Chen Z., Liu H., Zhang L., Gao P., Li D. (2011) Biosynthesis and structural characteristics of selenium nanoparticles by *Pseudomonas alcaliphila.* Colloids Surf B Biointerfaces 88(1):196–201.

Zieve R., Ansell P.J., Young T.W.K., Peterson P.J. (1985) Selenium volatilization by *Mortierella* species. Trans Br Mycol Soc 84:177–179.

CHAPTER 4

Removal of selenite from wastewater in a *Phanerochaete chrysosporium* pellet based fungal bioreactor

A modified version of this chapter was published as:

E.J. Espinosa-Ortiz, E.R. Rene, E.D. van Hullebusch, P.N.L. Lens (2015) Removal of selenite from wastewater in a *Phanerochaete chrysosporium* pellet based fungal bioreactor. International Biodeterioration and Biodegradation. 102:361–369.

Abstract

The performance of a novel fungal bioreactor system containing pellets of *Phanerochaete chrysosporium* was investigated in a continuously operated bioreactor for 41 days to remove selenite (SeO_3^{2-}) from synthetic wastewater. These fungal pellets were produced *in situ* under batch conditions during 4 days of incubation in the presence of SeO_3^{2-} (10 mg Se L^{-1}, 5 g glucose L^{-1}, pH-4.5). Subsequently, the system was continuously fed with SeO_3^{2-} at selenium and glucose loading rates of 10 mg Se $L^{-1}d^{-1}$ and 0.95 g glucose L^{-1} d^{-1}, respectively, and a hydraulic retention time of 24 h. After achieving steady-state removal profiles (8 days, ~70% removal from 10 mg Se $L^{-1}d^{-1}$), the biomass was partially removed, once every 4 days, in order to limit the excessive growth of the fungus. Afterwards, the fungal pelletized reactor was tested for its response to an increase in the SeO_3^{2-} loading rate from 10 to 20 mg Se $L^{-1}d^{-1}$. During this phase (8 days), although there was a declining trend in the removal of SeO_3^{2-}, the bioreactor showed good resilience to the doubled SeO_3^{2-} concentration. The bioreactor was further subjected to intermittent spikes of SeO_3^{2-} (30-50 mg Se L^{-1}) once every 4 days. The bioreactor showed a good adaptability and flexibility by recovering to every intermittent spike of SeO_3^{2-}, achieving ~70% total soluble Se removal from the continuous Se loading rate (10 mg Se $L^{-1}d^{-1}$). The presence of SeO_3^{2-} influenced the morphology of the fungal pellets, and assisted in controlling excess biomass growth. This study shows that fungal bioreactors can handle fluctuating loads of aqueous-phase SeO_3^{2-}, while simultaneously offering the possibility to synthesize elemental selenium under long-term operations.

Key words: Selenium removal, fungal pellets, up-flow bioreactor, *Phanerochaete chrysosporium*

4.1 Introduction

Selenium (Se) is an essential trace element for life, however, at high concentrations it can be highly toxic (Lenz and Lens, 2009). Se can be present in different oxidation states in the natural environment, and the soluble oxyanions selenate (SeO_4^{2-}) and selenite (SeO_3^{2-}) are the most common forms of this element. The presence of Se oxyanions in the environment is associated with mining, glass manufacturing, coal and metal production, and some agricultural activities. Se concentrations depend on the wastewater type, ranging from a few to thousands of $\mu g\ L^{-1}$ (Lenz and Lens, 2009). Therefore, studies pertaining to the adaptability and flexibility of Se removal systems are crucial for their full-scale application, where fluctuations or spikes of pollutants are quite common in the waste streams due to unexpected loss of influent flow, shutdowns or other transients naturally associated with the industrial or agricultural activity.

Different physicochemical methods have been used for the removal of Se oxyanions from water (Bleiman and Mishael, 2010; Geoffroy and Demopoulos, 2011; Zelmanov and Semiat, 2013). In the last decades, however, the use of biological agents for the treatment of Se polluted effluents has become popular, attracting the attention of scientists for the investigation of new Se-reducing organisms. Table 4.1 overviews the different bioreactors that have been used to treat wastewater polluted with Se. The ability of some microorganisms, particularly bacteria, to reduce Se oxyanions to elemental selenium (Se^0) is well known. Some fungal species have also shown to be capable of transforming SeO_3^{2-} to the less toxic, insoluble and more stable form Se^0 (Gharieb et al., 1995; Sarkar et al., 2011; Espinosa-Ortiz et al., 2015). The use of fungi over bacteria to remove Se from wastewater can be advantageous since fungi have a set of characteristics that allow them to grow under conditions that are less favorable for bacteria. Fungi can grow under acidic to neutral pH ranges (3.0-7.0). Of particular interest are the acid tolerant fungal strains, which are useful for treating acidic or mildly acidic effluents (*i.e.,* acid mine drainage and acid seeps). Moreover, fungi have the ability to grow at low moisture levels and metabolize organic compounds in low nitrogen and phosphorus environments (McKinney, 2004). Furthermore, these organisms are simple to cultivate, easy to handle at the laboratory scale and can be as well easily scaled-up. Fungi serves as an excellent biocatalyst for the biodegradation of organic pollutants and they are also a good biosorbent for removing heavy metals present in wastewaters (Say et al., 2001; Levin et al., 2003; Bayramoğlu and Arica 2008; Xu et al., 2012; Pakshirajan et al., 2013; Verma et al., 2013). However, little attention has been given to the use of fungal bioreactors, particularly fungal pellets.

The aim of this study was, therefore, to explore the performance of a novel fungal bioreactor inoculated with pellets of *Phanerochaete chrysosporium* for the continuous removal of SeO_3^{2-} from synthetic wastewater. This fungus is well known for its ability to degrade a wide range of organic pollutants (Denizli et al., 2004; Sedighi et al., 2009; Behnood et al., 2014) and has recently been described as a Se-reducing organism (Espinosa-Ortiz et al., 2015). The adaptability and resilience capacity of the fungal bioreactor was assessed by testing its response to high Se concentration loads and Se spikes. In order to better understand the fate of Se in the bioreactor, a simple mass balance was done in the aqueous, biomass and volatile phases of the fungal bioreactor. The development of the fungal biomass over time was also monitored during the entire operational time of 41 days.

Table 4.1 Bioreactors used for the treatment of Se polluted effluents.

Reactor type	Inoculum	$Se_{influent}$ (mg L^{-1})	Operational conditions	Removal efficiency	References
Chemostat reactor	*Bacillus* sp.	41.8	pH 7.8-8.0 30 °C HRT 95.2 h	99%	Fujita et al., 2002
UASB reactor	Anaerobic granular sludge	7.9	pH 7.0 30°C HRT 6 h methanogenic and sulfate reducing conditions	Methanogenic conditions 99% Sulfate reducing conditions 97%	Lenz et al., 2008
Column reactor	Sulfate-reducing bacteria	2	pH 6.3-8.2 18-22°C HRT 48 h	>95%	Luo et al., 2008
Suspended sludge bed reactor	Anaerobic granular sludge	1.5-3.5	pH 7.5 30°C HRT 24 h	60%	Soda et al., 2011
UASB reactor	Anaerobic granular sludge	1.5-3.5	pH 7.5 30°C HRT 24 h	>95%	Soda et al., 2011
Up-flow fungal pelleted reactor	*Phanerochaete chrysosporium*	10	pH 4.5 30°C HRT 24 h	70%	This study

4.2 Materials and methods

4.2.1 Strain, medium composition and pre-cultivation of fungal cultures

The white-rot fungus *Phanerochaete chrysosporium* MTCC187 (Institute of Microbial Technology, Chandigarh, India) was used in this study. The fungus was maintained on malt extract agar. A stock spore fungal suspension was obtained by harvesting fungi grown on agar plates for 3 days into sterilized water and then maintained at 4°C. Pre-cultivation cultures of the fungus were achieved by inoculating 2% (*v/v*) of the fungal spore solution into 50 mL of sterilized nitrogen-limited liquid medium (pH 4.5, 100 mL Erlenmeyer flasks) and incubated at 30°C in an orbital shaker at 150 rpm for 2 days. Sub-cultures of the 2 day old fungi were used as the inoculum for the reactor. The components of the nitrogen-limited medium were (g L^{-1}): glucose, 10; KH_2PO_4, 2; $MgSO_4 \cdot 7H_2O$, 0.5; NH_4Cl, 0.1; $CaCl_2 \cdot 2H_2O$, 0.1; thiamine, 0.001 and 5 mL of trace elements solution (Tien and Kirk, 1988).

4.2.2 Bioreactor configuration and operating conditions

Se removal experiments were performed in a bench-scale up-flow reactor (Figure 4.1), which consisted of a feed tank, a feed peristaltic pump, a flow-meter to regulate air supply, a glass reactor, an effluent tank and two gas traps containing 65% HNO_3 to recover the volatile fraction of Se

(Winkell et al., 2010). The reactor consisted of a 1 L cylindrical glass container with a working volume of 0.8 L (Table 4.2). The up-flow velocity in the reactor was 2.4 m h^{-1}, achieved from the liquid flow rate and no reactor liquid recirculation was applied. For start-up, the reactor was sterilized and filled with 0.8 L of sterile nitrogen-limited medium (pH 4.5). Sterile conditions were maintained throughout the experiment. Sub-cultured pellets were fragmented into pieces, homogenized and used as inoculum (0.77 g dry biomass L^{-1}). The actual up-flow reactor used in this study is shown in Figure 4.2.

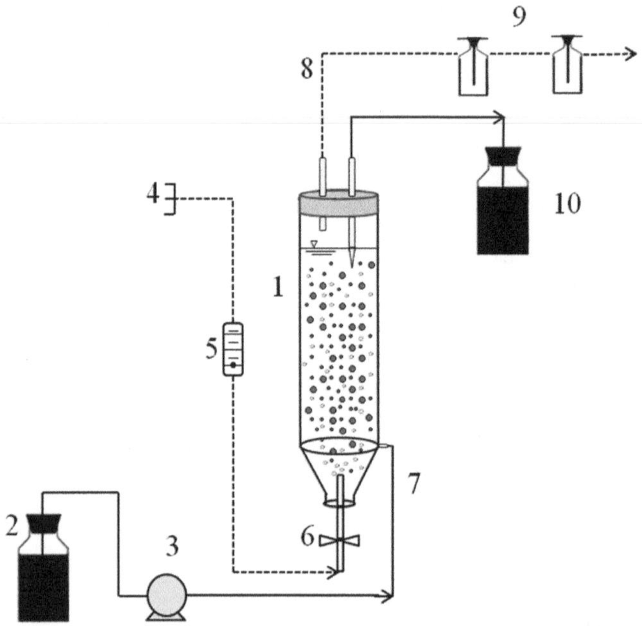

Figure 4.1 Fungal pelleted reactor. 1) Reactor; 2) Growth medium tank; 3) Peristaltic pump; 4) Air flow; 5) Flow-meter; 6) Valve; 7) Influent port; 8) Gas outlet; 9) Gas traps; 10) Effluent tank.

Table 4.2 Operational conditions of the fungal pelleted bioreactor.

Parameter	Value
Working volume	0.8 L
Air flow	0.156 vvm*
Up-flow velocity	2.4 m h^{-1}
Inoculum	0.77 g L^{-1}
Hydraulic retention time	24 h
Flow rate	0.033 L h^{-1}
Initial pH	4.5
Temperature	30°C

Note: * vvm = volume of air under standard conditions per volume of liquid per minute

Figure 4.2 Up-flow fungal pelleted reactor.

The operational parameters used for the bioreactor operation are summarized in Table 4.2. Different operational periods were used for the experimental run of the bioreactor. Descriptions of each operational period and the corresponding glucose and SeO_3^{2-} loading rates are depicted in Table 4.3 and Figure 4.3, respectively. The bioreactor was operated under batch conditions (period I) for 4 days in order to facilitate the formation of fungal pellets and expose the fungi to SeO_3^{2-} (10 mg Se L^{-1}). Subsequently, continuous operation of the reactor (period II) was performed between 4-25 days. After achieving steady-state removal profiles (8 days), biomass was partially removed to limit the excessive growth of the fungi, once every 4 days, until 16 days. The response of the bioreactor to an increase of the Se concentration was tested by maintaining 20 mg Se L^{-1} in the influent. Afterwards, intermittent spikes (period III) of SeO_3^{-2} (30 and 50 mg Se L^{-1}) were applied to the system once every 4 days, from days 25 to 41. The performance of the reactor was determined by estimating the glucose and Se removal rates ($R_{glucose}$ and $R_{selenium}$, respectively), according to Equations 1 and 2, for batch and continuous reactor, respectively. The Se removal efficiency was estimated according to Equation 3.

Batch reactor:

$$Removal\ rate, R\ (mg\ L^{-1}d^{-1}) = \frac{Initial\ concentration\ (mg\ L^{-1}) - Concentration\ at\ t\ (mg\ L^{-1})}{Different\ time\ steps\ (d)} \quad \text{Eq. (1)}$$

Continuous reactor:

$$Removal\ rate, R\ (mg\ L^{-1}d^{-1}) = Flow(L\ d^{-1}) \left(\frac{Initial\ concentration\ (mg\ L^{-1}) - Concentration\ at\ t\ (mg\ L^{-1})}{Volume\ in\ the\ reactor\ (L)}\right) \quad \text{Eq. (2)}$$

$$Removal\ efficiency, (\%) = \left(\frac{Initial\ concentration\ (mg\ L^{-1}) - Concentration\ at\ t\ (mg\ L^{-1})}{Initial\ concentration\ (mg\ L^{-1})}\right) \times 100 \quad \text{Eq. (3)}$$

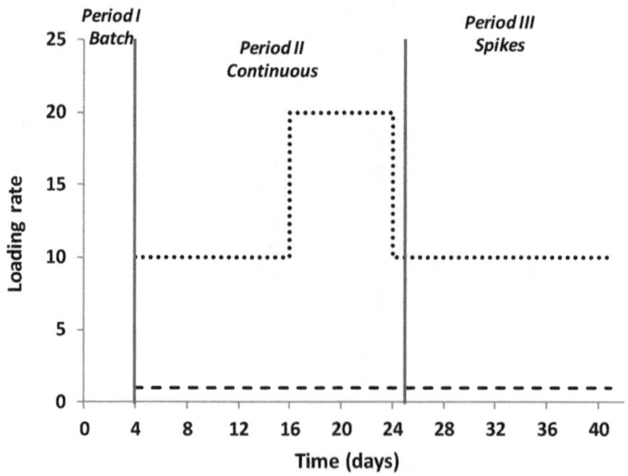

Figure 4.3 Loading rates for different operational periods during the experimental run of the bioreactor. ··· SeO_3^{2-} (mg Se $L^{-1}d^{-1}$) – – Glucose (g L^{-1} d^{-1}).

Table 4.3 Description of the operating periods of the fungal pelleted reactor.

Operation time (days)	Operating period	Description
0-4	I. Batch	Formation of the fungal pellets. Initial glucose concentration 5 g L^{-1}
4-8	II. Continuous	Continuous supply of growth media and Se
8-12	Removal of biomass	Removal of biomass - 0.10 g dry weight
12-16	Removal of biomass	Removal of biomass - 0.09 g dry weight
16-24	Increase of Se concentration	Doubled Se load rate
24-25	Maintenance	Depuration of the excessive Se concentration
25-33	III. Se spike	Se spike-30 mg Se L^{-1} supplied once every 4 days
33-41	Se spike	Se spike-50 mg Se L^{-1} supplied once every 4 days

4.2.3 Analytical methods

Samples were withdrawn from the inlet and outlet of the reactor twice per day. The samples were centrifuged at 37,000 × g for 15 min, and the supernatant was then used for analysis. The glucose concentration was analyzed with the dinitrosalicylic acid method using D-glucose as standard (Miller, 1959). Samples were filtered (0.45 µm, Type GF/F, Whatman Inc., Florham Park, NJ) and preserved with acidified solution of 0.5% HNO_3 in ultrapure water (Milli-Q water, 18MΩ-cm) for the measurement of total soluble Se with ICP-MS as described previously (Espinosa-Ortiz et al., 2015). The volatile fraction of Se was recovered from the 65% HNO_3 gas traps, which were sampled and replaced with new acid solution after each operating period.

Gravimetric determination of the dry biomass was done by filtering the biomass suspension through a pre-dried (24 h at 105°C) and pre-weighted filter paper (Type GF/F, Whatman Inc., Florham Park, NJ) of 0.45 µm pore-size. In order to quantify the Se uptake by the fungi, the total Se contained in the final fungal dry biomass was estimated after a microwave-assisted (CEM corporation, model MARS, California, USA) acid digestion with 10 mL of HNO_3 65% (Method 3030K, APHA, 2005).

A series of digital images were captured (Canon EOS Rebel T3, Taiwan) in order to visualize the morphological development of the fungal biomass during different time durations of the continuous experiment. The size of the fungal pellets was estimated using a Vernier caliper. To determine the settling and dewatering properties of the fungal pellets, the sludge volume index (SVI) and capillary suction time (CST) were analyzed according to standard methods (2710D, 2710G, APHA, 2005). The CST was determined using a CST test kit (Triton Electronics, model 200, Essex, UK), with CST filter paper (Triton Electronics) and a 18 mm sludge reservoir.

Samples were taken from the effluent once every three days to account for bacterial contamination. The samples were serially diluted and plated in Petri dishes containing plate count agar. After 24 h of incubation, at 30 °C, bacterial counting was done using a colony counter (UIL Instruments, Spain), as described by Safont et al. (2012).

4.2.4 Statistical analysis

The maximum substrate removal and Se removal rates are reported in this study. The mean and standard deviation values were calculated for Se removal efficiency when steady state conditions were attained for each operating period. The effect caused by the operating period on the removal efficiency was evaluated by performing analysis of variance (ANOVA) at $P \leq 0.05$. The Tukey Test at $P \leq 0.05$ was applied whenever results showed significant differences.

4.3 Results

4.3.1 Bioreactor operation in batch mode

In order to allow the formation and growth of the fungal pellets and to acclimate the fungus to SeO_3^{2-}, the bioreactor was started in batch mode (period I) with a medium containing 10 mg Se L^{-1} and 5 g glucose L^{-1} as the substrate. This start-up period lasted for 4 days. The initial pH was fixed at 4.5, but dropped to around 3.0 after 1 day of operation, maintaining this value throughout this operational period (Figure 4.4A). Figure 4.4B shows that glucose was constantly consumed, indicating that *P. chrysosporium* was active during this start-up period, with a maximum substrate removal rate ($R_{glucose}$) of 0.38 g glucose $L^{-1}d^{-1}$. The maximal Se removal rate ($R_{Selenium}$) in batch mode was ~2.3 mg $L^{-1}d^{-1}$, with an average removal efficiency of 46.5 ± 0.7%. The removal of Se during the batch mode is depicted in Figure 4.4C. The volatile fraction of Se during this mode of bioreactor operation was estimated to be less that 5% of the total Se removed. The performance of the reactor under different operating periods showed to be statistically significant with a *P* value of 7.5×10^{-9}. The removal efficiencies observed during the batch period was similar ($P \geq 0.05$) to those obtained during the first removal of biomass, the first shock load (20 mg L^{-1}) and the first Se spike (30 mg Se L^{-1}) experiments performed in this bioreactor.

4.3.2 Continuous bioreactor operation at constant Se concentrations

After the batch operational mode, the bioreactor was operated in continuous mode (period II) and the reactor was fed at an organic loading rate of 0.95 g glucose $L^{-1}d^{-1}$ and a SeO_3^{2-} loading rate of 10 mg Se $L^{-1}d^{-1}$. The reactor was operated at a HRT of 1 day, for 4 days. The pH of the feeding solution was 4.5, but within the reactor the pH dropped to 3.5, which was nearly constant throughout this operational period (Figure 4.4A). Removal of glucose in the reactor during the continuous phase is shown in Figure 4.4B. The maximum substrate removal rate was 0.44 g glucose $L^{-1}d^{-1}$. The average Se removal efficiency at steady-state after 4 days of continuous operation was about 67.3 ± 5.1% ($R_{Selenium}$ 7.3 mg Se $L^{-1}d^{-1}$). The removal efficiency during the first 4 days of continuous bioreactor operation was significantly higher compared to the values observed during batch period ($P = 1 \times 10^{-3}$). The performance of the reactor during this period was similar to those obtained during the supply of the second spike of 30 mg Se L^{-1} and the spikes of 50 mg Se L^{-1} ($P \geq 0.05$).

In order to control the overgrowth of the fungal biomass, a maintenance step of removing biomass was introduced (Table 4.1). The reactor was intermittently stopped, the medium was decanted and then the reactor was filled again with fresh medium. This strategy to control the biomass growth was repeated twice during operational period II, once every 4 days. The starting pH value after the removal of biomass was 4.5, which decreased to 3.5 (Figure 4.4A). Maximum substrate removal rates were 0.4 and 0.5 g glucose $L^{-1}d^{-1}$ after the first and second steps of biomass removal, respectively. It was observed that the removal of Se considerably decreased after removing biomass from the reactor, with $R_{selenium}$ and removal efficiency values of 5.2 mg Se $L^{-1}d^{-1}$ and 44.8 ± 3.2%, respectively, after the first removal step, and 3.0 mg Se $L^{-1}d^{-1}$ and 28.7 ± 0.7%, respectively, after the second removal step.

The system was evaluated against a constant increase in the Se concentration, which was twice the former operational load (20 mg Se $L^{-1}d^{-1}$), for about 8 days. The maximum substrate removal rate during the first 4 days of operation under increased Se conditions was 0.48 g glucose $L^{-1}d^{-1}$, with a maximum Se removal efficiency of 38.9 ± 0.4% ($R_{selenium}$ 7.4 mg Se $L^{-1}d^{-1}$). During the last 4 days of operation, the $R_{glucose}$ was 0.42 g $L^{-1}d^{-1}$, and a clear decrease in the removal efficiency of Se was observed (19 ± 1.2%) with a $R_{selenium}$ value of 4.1 mg Se $L^{-1}d^{-1}$. The removal efficiency of the reactor was found to be the lowest during the last 4 days of the shock loading period (20 mg L^{-1}). Due to the low removal efficiency achieved after the increase of the Se concentration, a maintenance step was performed. For 24 h, the reactor was fed at a Se loading rate of 10 mg $L^{-1}d^{-1}$, in order to dilute the incoming feed and also to ensure favorable conditions that could prevent Se induced toxic effects on the fungal pellets.

The biomass concentration at the end of the experimental run, without taking into account the biomass that was removed from the system during days 16 to 41 was 1.4 g dry biomass L^{-1}. The total Se accumulated by the fungal biomass was estimated to be 71.2 mg Se g^{-1} dry biomass, which accounted for ~84% of the total Se removed in the bioreactor. During the continuous operation of the reactor, the volatile fraction of Se averaged at 9.3 ± 0.6% of the total Se removed.

4.3.3 Bioreactor response to spikes of Se concentration

The overall response of the fungal pelleted reactor to significant fluctuations (period III) of Se concentrations in the influent was tested in order to simulate conditions that are frequently encountered during industrial operations. The pH in the reactor decreased from 4.5 to around 3.0 after each spike of Se (Figure 4.4A).

The removals of glucose and Se in the bioreactor during period III are shown in Figures 4.4B and 4.4C, respectively. After the first spike (30 mg Se L^{-1}), the $R_{glucose}$ was 0.52 g $L^{-1}d^{-1}$, and the maximum Se removal efficiency was 48.5 ± 0.2% ($R_{selenium}$ 4.8 mg Se $L^{-1}d^{-1}$). After the second spike (30 mg Se L^{-1}), the maximum substrate removal rate increased to about 0.80 g $L^{-1}d^{-1}$, and during this spike the removal efficiency increased to 68.7 ± 0.4% ($R_{selenium}$ 6.9 mg Se $L^{-1}d^{-1}$). During the third spike (50 mg Se L^{-1}), $R_{glucose}$ was 0.74 g $L^{-1}d^{-1}$ and the maximum Se removal rate was 6.6 mg Se $L^{-1}d^{-1}$, corresponding to a removal efficiency value of 67.3 ± 0.5%. At the end of the fourth spike (50 mg Se L^{-1}), the maximum substrate removal rate was 0.82 g $L^{-1}d^{-1}$, with a maximum Se removal efficiency of 69 ± 1.5% ($R_{selenium}$ 6.5 mg Se $L^{-1}d^{-1}$).

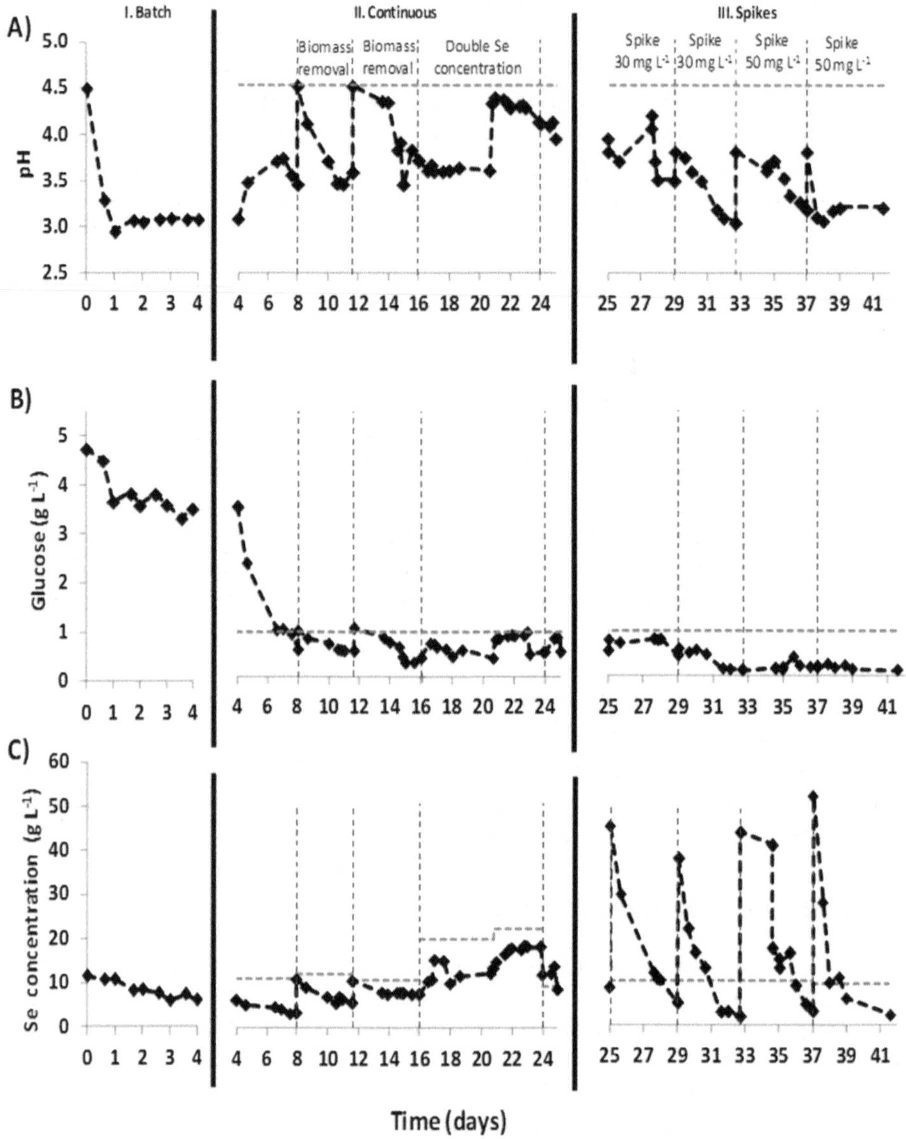

Figure 4.4 Time-course profile in the fungal bioreactor for A) pH, B) Glucose consumption, C) Total soluble Se removal. --- Influent -♦- Effluent --- Boundary between different actions taken during the operating period.

4.3.4 Evolution and growth of the fungal biomass

The evolution of the fungal biomass in the reactor was examined based on a series of photographs taken during the different operational periods of the bioreactor (Figure 4.5). The first 4 days of operation in batch mode allowed the formation of the fungal pellets. On day 1 the biomass consisted of small, fluffy and white-ivory pellets which developed into more compact and spherical mature pellets. At the end of day 4, the pellets turned to a red-orange color, which indicated the presence of Se^0 (Gharieb et al., 1995; Sarkar et al., 2011). During the first few days of operation of the bioreactor, the biomass morphology had a mixture of dense and spherical hyphae-free pellets and some suspended non-uniform clumps (Figure 4.5). The size of the fungal pellets was not homogeneous throughout the different operating periods of the reactor. Removed biomass from the bioreactor between days 8 and 16 varied between 3 ± 0.5 mm and 10 ± 0.3 mm in diameter. During this period (between 8 and16 days), the CST was in average 5.6 s; whereas the SVI amounted to 52 ± 8 mL g^{-1}. Typical CST values for sludge solids less than 20 s represent good dewaterability (Subramanian et al., 2008, Triton electronics, 1998), whereas low SVI values (50-150 mL g^{-1}) are related to fast settleability (Templeton and Butler, 2011).

After day 16, when the concentration of Se was doubled (20 mg L^{-1}), there was an increase in the number of pellets with a compact and spherical appearance, and no visible growth of biomass was observed in the reactor during this period (Figure 4.5). After 35 days of operation, the formation of hyphal branches and aggregates of pellets was observed, provoking their breakdown, which in turn caused the increase of viscosity in the bioreactor due to the presence of dispersed mycelia. No sign of bacterial contamination was observed throughout the experiment until the reactor was stopped. The dilution series of the effluent samples that were plated and incubated at 30 °C did not yield any bacterial colonies.

4.4 Discussion

4.4.1 Removal of SeO_3^{2-} in a fungal pelleted bioreactor

This study envisaged the performance of an up-flow fungal bioreactor containing pellets of *P. chrysosporium* to remove SeO_3^{2-} from synthetic wastewater under different operational modes and experimental conditions. The system showed better removal efficiency during the continuous operation compared to the batch operating period. The bioreactor did not show a favorable response to a higher Se loading rate (20 mg Se $L^{-1}d^{-1}$), decreasing the removal efficiency to ~19% after 8 days of continuous operation under these conditions (Figure 4.4C). Despite a declining trend in the removal of total soluble Se, the system showed good resilience by being able to achieve up to ~70% removal efficiency after the Se loading rate was set back to 10 mg Se L^{-1} and when Se spikes were added to the reactor.

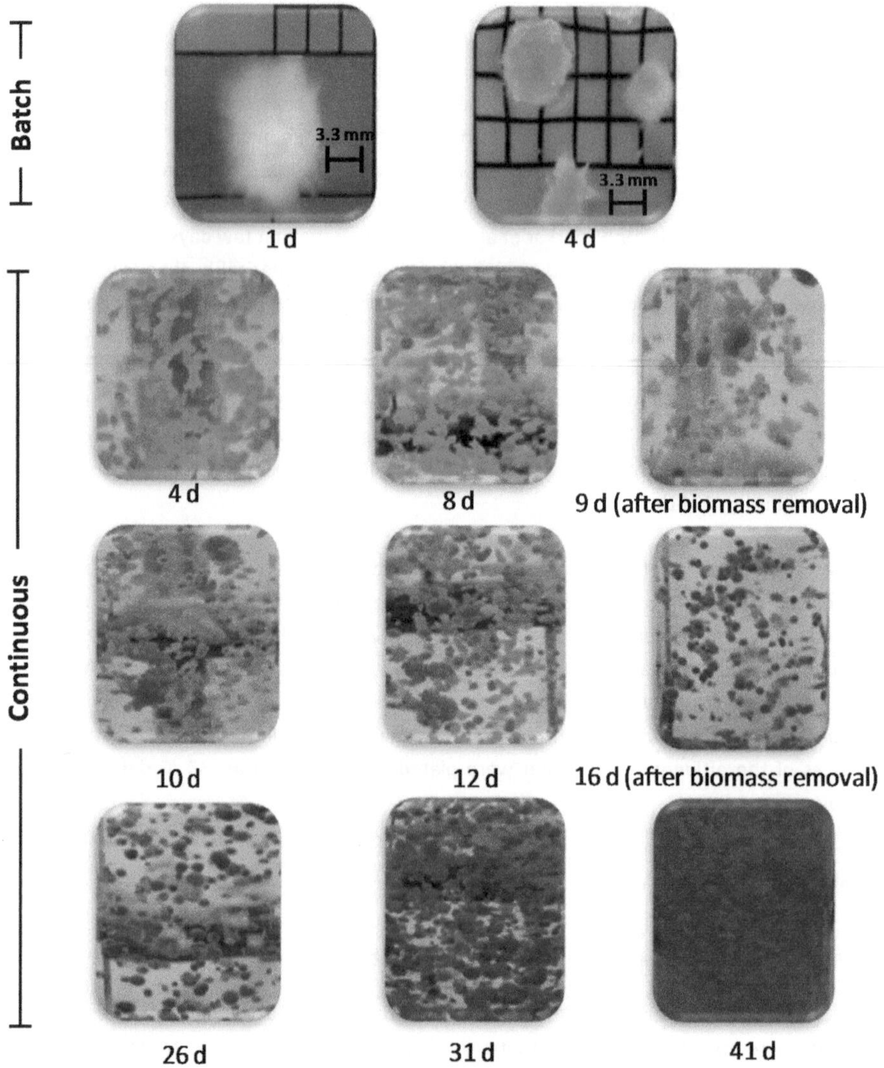

Figure 4.5 Morphological development of the fungal biomass in the bioreactor over 41 days of continuous operation.

The removal efficiency of Se obtained under steady-state conditions (~70% from 10 mg Se L^{-1} d^{-1}) might appear low when compared to the removal efficiencies achieved in other biological treatment systems, particularly those using bacteria, wherein removal efficiencies often exceed 90% (Table 4.1). However, it should be highlighted that in those studies a pre-conditioning of the wastewater was a prerequisite for the optimal performance of the reactor working at nearly neutral-basic conditions (pH 6.0-8.0), whereas this study was performed under acidic conditions (pH 4.5). The presence of volatile fractions of Se was also estimated in this study, accounting for less than 5% during the batch mode of operation (Period I) and ~10% of the total Se removed during the continuous mode of operation (Period II). Biomethylation of Se compounds is a well known fungal mechanism which is usually based on the transformation of inorganic Se forms to volatile derivates, most commonly as dimethylselenides (Gadd, 1993). The observation of Se volatile fraction in the batch and continuous modes of bioreactor operations suggests that the volatilization process is associated to the growth and non-growth fungal phases, as reported previously by Brady et al. (1996). The total Se in the fungal biomass accounted for ~84% of the total Se removed in the bioreactor. As demonstrated in a previous study, *P. chrysosporium* is able to reduce SeO_3^{2-} to Se^0 intracellularly in the form of nanoparticles (30-400 nm) (Espinosa-Ortiz et al., 2015). This clearly suggests that all the Se contained within the fungal biomass in this study could possibly correspond to Se^0 nanoparticles. Briefly stating, considering the simple mass balance calculations done in this study, 84% Se was removed by the biomass and 10% Se was removed as volatile fractions, while 6% of the total Se fed to the reactor during the continuous period remained unaccounted. This unaccounted Se fraction can be attributed to an incomplete trapping of the volatile Se fraction of Se (Brady et al., 1996).

The start-up period of the bioreactor was initiated with 5 g L^{-1} of glucose as the substrate. Typically, concentrations as high as 10 g L^{-1} are used to grow *P. chrysosporium* pellets (Moreira et al., 1996; Chen and Ding, 2012). This substrate and concentration were applied in this study as to achieve a proof of concept of the use of *P. chrysosporium* pellet fungal reactors to treat Se rich wastewaters. Further research is, however, needed to optimize the electron donor type and its concentrations in start-up and continuous operation of fungal pelleted reactors treating rich SeO_3^{2-} wastewaters.

Different fungal bioreactors have been proposed for wastewater treatment, the most common ones are the stirred tank reactor, bubble column reactor, fluidized bed reactor and airlift reactor. These bioreactor configurations have shown to be robust and efficient in treating a wide range of pollutants under well-controlled laboratory conditions, including dyes (Hai et al., 2013), endocrine disrupting contaminants (Blánquez and Guiesse, 2008), heavy metals (Pakshirajan and Swaminathan, 2009; Verma et al., 2013), pharmaceutical compounds (Rodarte-Morales et al., 2012) and phenolic compounds (Farkas et al., 2013), among others. Much of the technology developed is, however, still in its infancy for applications using fungi for the continuous treatment of effluents. There is thus a need for further research that will enhance our understanding on the long-term performance of mycoreactors, process intensification by preventing bacterial contamination and process optimization that facilitates the recovery and reuse of the fungal biomass.

4.4.2 Response of the system to Se spikes

Fluctuations or spikes in the concentration of different pollutants in wastewater are quite common in industrial situations and for real applications, the reactor stability against high concentrations of pollutants should always be considered in designing a reactor system (Tandukar et al., 2006). Therefore, it is crucial to determine the adaptability and flexibility of pollution control systems. Significant variation in the speciation and concentration of Se exists among different industrial or agricultural activities; even high variations might occur within the same wastewater treatment facilities over time. Fungal pelleted systems have shown to be able to tolerate and recover from fluctuations of pollutant concentrations. Sanghi et al. (2011) showed the ability of fungal pellets of *Coriolus versicolor* to tolerate high concentrations of toxic sulfonic azo dyes in sequential batch mode, sustaining its decolorizing potential for four continuous cycles. In this study, the fungal pelleted bioreactor was subjected to intermittent spikes of SeO_3^{2-} (30-50 mg Se L^{-1}, period III). The fungal system showed good adaptability and flexibility by recovering to every intermittent spike of SeO_3^{2-}. After the first spike of SeO_3^{2-} (either 30 or 50 mg Se L^{-1}), it took about 3 days for the system to recover, whereas after the second spike, the system was able to recover after 2 days, achieving about 70% soluble Se removal (Figure 4.4C).

4.4.3 Fungal morphology in the bioreactor

The up-flow bioreactor configuration and up-flow velocity (2.4 m h^{-1}) used in this study favored the fungal growth in an appropriate morphology as pellets, avoiding some of the most common operational difficulties when dispersed filaments are present in a bioreactor, such as an increase in the viscosity of the medium, decrease of oxygen supply and wrapping of the dispersed filaments around agitators, baffles or other internal elements of the reactors. Different factors have been suggested to favor the fungal pelletization process, including the use of carrier materials or additives such as surfactants (Callow and Ju, 2012). Due to their improved mechanical strength, rigidity and porosity characteristics, the use of pellets in bioreactors provides advantages over dispersed mycelium, making the system more resistant to environmental fluxes, allowing regeneration, reuse and easy separation of the biomass, as well as minimal clogging of the system (Braun and Vecht-Lifsitz, 1991; Renganathan et al., 2006; Sumathi and Manju, 2000; Xin et al., 2012).

In this study, an evolution of the fungal biomass morphology in the reactor over time was clearly observed. The presence of Se certainly had an effect on the fungal morphology of *P. chrysosporium*, promoting to have denser, compact and smother pellets, with barely noticeable fluffy zones with higher concentrations of Se (Figure 4.5). Towards the end of the experiment (35 days) fungal pellets deteriorated, showing filamentous growth, aggregation and breakdown of pellets. For this reason the reactor operation was terminated despite the good Se-reduction achieved towards the end of this study (about 70%). Strategies to avoid the deterioration would be to remove biomass or increase again the Se loading rate in the reactor.

4.4.4 Operational advantages of fungal pelleted reactors

This study shows that fungal pellets were resistant to variable SeO_3^{2-} fluxes and spikes, maintaining a reasonable SeO_3^{2-} removal efficiency (~70% from 10 mg Se $L^{-1}d^{-1}$). The use of different strategies (*i.e.,* biomass removal, maintaining acidic conditions) to prolong the longevity of the bioreactor was also demonstrated. Interestingly, the SeO_3^{2-} concentration itself can be used as a parameter to control the fungal biomass growth and morphology. Moreover, this study shows the use of fungal pellets as a good recovery method to capture the reduced elemental Se within the fungal biomass, enabling its easy separation from the treated effluent.

Even though in this study the reactor was run under sterile conditions, there is always the possibility of bacterial colonization. No significant bacterial contamination was observed for the time that the reactor was run. The acidic condition (pH 4.5-3.0) at which the reactor was run in this study could also be a good strategy to avoid bacterial growth, since most aerobic bacteria preferentially grow in neutral environments. As an example of the competition between bacteria and fungi in acidic environments, Rousk et al. (2009) showed that there was 30-fold increase in fungal importance, based on fungal growth/bacterial growth ratio, measured in a continuous soil pH gradient from 8.3 to pH 4.5.

Other studies have shown that even under non-sterile conditions, the reduction of medium pH in fungal reactors decreases the probability of bacterial proliferation (Libra et al., 2003).

The use of fungal pelleted reactors allows the possibility to recover the synthesized material (Se^0) and to reuse the fungal biomass. In this study, a strategy similar to a sequencing batch reactor (fill, react, settle and decant step) was followed in order to remove the excess biomass (during 8-16 days of operation) from the bioreactor. The good dewaterability and settleability property shown by the fungal pellets in this study (Section 4.3.4) facilitates the easy separation of the fungal biomass and Se^0 from the liquid phase. Entrapment of Se^0 in the fungal biomass avoids the use of centrifugation or filtration during the recovery step (Fujita et al., 2002). However, the development of an efficient protocol for the extraction and purification of Se^0 from the fungal biomass requires further investigation. Under laboratory scale conditions, Sonkusre et al. (2014) developed a procedure to recover pure and clean intracellular Se^0 nanoparticles from bacteria by achieving complete bacterial cell lysis through the use of lysozyme (enzyme used to break down bacterial cell walls) and a French press. A similar approach using an effective agent that is capable of breaking down the fungal cell walls could be used to recover Se^0 from the fungal cells.

4.4.5 Longevity of reactor operation

Maintaining the longevity of fungal reactors is quite challenging; different factors influence the performance over the operational time of the system, including the ability of the fungal pellets to acclimatize to diverse conditions in the reactor, the pollutant load applied and the possibility of bacterial contamination (Bochert and Libra, 2001). In this study, the bioreactor was operated continuously for 41 days, and the reactor performance was maintained effectively working for 35 days (maximal biomass retention time) before any sign of dispersed filamentous growth was observed in the reactor. On average, most fungal pelleted reactors are able to effectively run for about 30 days of operation, although there are a few studies that show working fungal pelleted reactors for even more than 3 months under sterile conditions (Borchert and Libra, 2001; Blánquez and Guieysse, 2008). In this study, it was observed that operating with high SeO_3^{2-} loading rates (20 mg Se L^{-1} d^{-1}) prevented the overgrowth of the fungal biomass (Figure 4.5). Thus, applying high SeO_3^{2-} loading rates could be used as a strategy to control fungal growth and procure the longevity of the bioreactors.

The reactor was intermittently stopped to manually remove the excess biomass, which avoided having an overgrowth of fungal biomass while allowing to recover the produced elemental Se entrapped within the pellets. The removal of biomass from the bioreactors is a common practice to avoid rapid accumulation of biomass, which otherwise would lead to biomass overgrowth and clogging of the reactor. Strategies to prevent biomass accumulation in a bioreactor include physical, chemical and biological methods. The mechanical or manual removal of biomass is a very usual physical method to control the biomass production in bioreactors (Wübker et al., 1997), although one of its main drawbacks is the fact that it requires complex reactor design, which might increase the operational costs (Deshusses et al., 1998). Moreover, in this study, it is clearly evident that there is a decrease in the performance of the reactor in terms of Se removal efficiency. After removing biomass from the reactor (end of operating period III and IV), there was a reduction in the removal efficiency of Se by 30-40% due to the loss of active biomass in the reactor. While using the physical removal of biomass as control of fungal overgrowth, there should be equilibrium between the controlling of excess biomass and maintaining optimal pollutant removal efficiency.

4.5 Conclusions

This study demonstrated that *Phanerochaete chrysosporium* is capable of removing soluble Se (SeO_3^{2-}) from aqueous medium (pH 4.5) in an up-flow reactor operated using fungal pellets in batch and continuous mode under different operational conditions. During steady-state operations, the bioreactor achieved ~70% SeO_3^{2-} removal at a Se loading rate of 10 mg Se $L^{-1}d^{-1}$. The system showed good adaptability and flexibility to intermittent spikes of SeO_3^{2-} (30 and 50 mg L^{-1}). The bioreactor was operated continuously for 41 days, and was maintained to work efficiently for 35 days before any sign of dispersed filamentous growth was noticed in the reactor. Due to the good settleability of the fungal pellets and the entrapment of the Se^0 within the fungal biomass, this bioreactor configuration facilitates the recovery of the Se^0 for further use.

4.6 References

APHA (1995) Standard Methods for Water and Wastewater Examination. 19th ed. American Public Health Association, Washington, DC, USA.

APHA (2005) Standard methods for examination of water and wastewater. 20 ed. American Public Health Association. Washington, DC, USA.

Behnood M., Nasernejad B., Nikazar M. (2014) Biodegradation of crude oil from saline waste water using white-rot fungus *Phanerochaete chrysosporium*. J Ind Eng Chem 20:1879–1885.

Blánquez P., Guieysse B. (2008) Continuous biodegradation of 17β-estradiol and 17α-ethynylestradiol by *Trametes versicolor*. J Hazard Mater 150:459–462.

Bleiman N., Mishael Y.G. (2010) Selenium removal from drinking water by adsorption to chitosan-clay composites and oxides: Batch and columns tests. J Hazard Mat 183:590–595.

Borchert M., Libra J.A. (2001) Decolorization of reactive dyes by the white rot fungus *Trametes versicolor* in sequencing batch reactors. Biotechnol Bioeng 75:313–321.

Braun S., Vecht-Lifsitz S.E. (1991) Mycelial morphology, and metabolite production. Trends Biotechnol 9:63–68.

Brady J.M., Tobin J.M., Gadd G.M. (1996) Volatilization of selenite in aqueous medium by a *Penicillium* species. Mycological Research 100:955–961.

Bayramoğlu G., Arıca M.Y. (2008) Removal of heavy mercury(II), cadmium(II) and zinc(II) metal ions by live and heat inactivated *Lentinus edodes* pellets. Chem Eng J 143:133–140.

Callow N.V., Ju L.K. (2012) Promoting pellet growth of *Trichoderma reesei* Rut C30 by surfactants for easy separation and enhanced cellulase production. Enzyme Microb Technol 50:311–317.

Chen B., Ding J. (2012) Biosorption and biodegradation of phenanthrene and pyrene in sterilized and unsterilized soil slurry systems stimulated by *Phanerochaete chrysosporium*. J Hazard Mater 229-230:159–169.

Denizli A., Cihangir N., Rad A.Y., Taner M., Alsancak G. (2004) Removal of chlorophenols from synthetic solutions using *Phanerochaete chrysosporium*. Process Biochem 39:2025–2030.

Deshusses M.A., Cox H.H.J., Miller D.W. (1998) The use of CAT scanning to characterize bioreactors for waste air treatment, Paper 98-TA20B.04. In: Proceedings of the Annual Meeting and Exhibition of the Air and Waste Management Association, June 15-17, San Diego, CA.

Espinosa-Ortiz E.J., Gonzalez-Gil G., Saikaly P.E., van Hullebusch E.D., Lens P.N.L. (2015) Effects of selenium oxyanions on the white-rot fungus *Phanerochaete chrysosporium*. Appl Microbiol Biotechnol 99:2405–2418.

Farkas V., Felinger A., Hegedüsova A., Dékány I., Pernyeszi T. (2013) Comparative study of the kinetics and equilibrium of phenol biosorption on immobilized white-rot fungus *Phanerochaete chrysosporium* from aqueous solution. Colloid Surf B: Biointerfaces 103:381–390.

Fujita M., Ike M., Kashiwa M., Hashimoto R., Soda S. (2002) Laboratory-scale continuous reactor for soluble selenium removal using selenate-reducing bacterium, *Bacillus* sp SF-1. Biotechnol Bioeng 80:755–761.

Gadd G.M. (1993) Microbial formation and transformation of organometallic and organometalloid compounds. FEMS Microbiol Rev 11:297–316.

Gharieb M.M., Wilkinson S.C., Gadd G.M. (1995) Reduction of selenium oxyanions by unicellular, polymorphic and filamentous fungi: cellular location of reduced selenium and implication for tolerance. J Ind Microbiol 14:300–311.

Geoffroy N., Demopoulos G.P. (2011) The elimination of selenium (IV) from aqueous solution by precipitation with sodium sulfide. J Hazard Mater 185:148–154.

Hai F.I., Yamamoto K., Nakajima F., Fukushi K., Nghiem L.D., Price W.E., Jin B. (2013) Degradation of azo dye Acid Orange 7 in a membrane bioreactor by pellets and attached growth of *Coriolus versicolour*. Bioresour Technol 141:29–34.

Lenz M., van Hullebusch E.D., Hommes G., Corvini P.F.X., Lens P.N.L. (2008) Selenate removal in methanogenic and sulfate-reducing up-flow anaerobic sludge bed reactors. Water Res 42:2184–2194.

Lenz M., Lens P.N.L. (2009) The essential toxin: The changing perception of selenium in environmental sciences. Sci Total Environ 407:3620–3633.

Levin L., Viale A., Forchiassin A. (2003) Degradation of organic pollutants by the white rot basiomycete *Trametes trogii*. Int. Biodeterior. Biodegradation 52:1–5.

Luo Q., Tsukamoto T.K., Zamzow K.L., Miller G.C. (2008) Arsenic, selenium and sulfate removal using an ethanol-enhanced sulfate-reducing bioreactor. Mine Water Environ 27:100–108.

Libra J.A., Borchert M., Banit S. (2003). Competition strategies for the decolorization of a textile-reactive dye with the white-rot fungi *Trametes versicolor* under non-sterile conditions. Biotechnol Bioeng 82:736–744.

McKinney R.E. (2004) Environmental pollution control microbiology: a fifty year perspective. CRC Press, NY, USA.

Miller G.L. (1959) Use of dinitrosalicylic acid reagent for determination of reducing sugar. Anal Chem 31:426–428.

Moreira M.T., Sanromán A., Feijoo G., Lema J.M. (1996). Control of pellet morphology of filamentous fungi in fluidized bed bioreactors by means of a pulsing flow. Application to *Aspergillus niger* and *Phanerochaete chrysosporium*. Enzyme Microbial Technol 19:261–266.

Pakshirajan K., Swaminathan T. (2009) Biosorption of copper and cadmium in packed bed columns with live immobilized fungal biomass of *Phanerochaete chrysosporium*. Appl Biochem Biotechnol 157:159–173.

Pakshirajan K., Izquierdo M., Lens P.N.L. (2013) Arsenic (III) removal at low concentrations by biosorption using *Phanerochaete chrysosporium* pellets. Sep Sci Technol 48:1111–1122.

Renganathan S., Thilagaraj W.R., Miranda L.R., Gautam P., Velan M. (2006) Accumulation of Acid Orange 7, Acid Red 18 and Reactive Black 5 by growing *Schizophyllum commune*. Bioresour Technol 97:2189–2193.

Rodarte-Morales A.I., Feijoo G., Moreira M.T., Lema J.M. (2012) Biotransformation of three pharmaceutical active compounds by the fungus *Phanerochaete chrysosporium* in a fed batch stirred reactor under air and oxygen supply. Biodegradation 23:145–156.

Rousk J., Brookes P.C., Bååth E. (2009) Contrasting soil pH effects on fungal and bacterial growth suggests functional redundancy in carbon mineralization. Appl Environ Microbiol 75:1589–1596.

Safont B., Vitas A.I., Peñas F.J. (2012) Isolation and characterization of phenol degrading bacteria immobilized onto cyclodextrin-hydrogel particles within a draft tube spouted bed bioreactor. Biochem Eng J 64:69–75.

Sanghi R., Dixit A., Verma P. (2011) Evaluation of *Coriolus versicolor* for its tolerance towards toxic sulphonic azo dyes in sequential batch mode. Process Saf Environ 89:15–21.

Sarkar J., Dey P., Saha S., Acharya K. (2011) Mycosynthesis of selenium nanoparticles. Micro Nano Lett 6:599–602.

Say R., Denizli A., Arica M.Y. (2001) Biosorption of cadmium(II), lead(II) and copper(II) with the filamentous fungus *Phanerochaete chrysosporium*. Bioresour Technol 76:67–70.

Sedighi M., Karimi A., Vahabzadeh F. (2009) Involvement of ligninolytic enzymes of *Phanerochaete chrysosporium* in treating the textile effluent containing Astrazon Red FBL in a packed-bed bioreactor. J Hazard Mater 169:88–93.

Soda S., Kashiwa M., Kagami T., Kuroda M., Yamashita M., Ike M. (2011) Laboratory-scale bioreactors for soluble selenium removal from selenium refinery wastewater using anaerobic sludge. Desalination 279:433–438.

Sonkusre P., Nanduri R., Gupta P., Singh Cameotra S. (2014) Improved extraction of intracellular biogenic selenium nanoparticles and their specificity for cancer chemoprevention. J Nanomed Nanotechnol 5:1–9.

Subramanian B.S., Yan S., Tyagi R.D., Surampalli R.Y. (2008) A new, pellet-forming fungal strain: its isolation, molecular identification, and performance for simultaneous sludge-solids reduction, flocculation, and dewatering. Water Environ Res 80:840–852.

Sumathi S., Manju B.S. (2000) Uptake of reactive textile dyes by *Aspergillus foetidus*. Enzyme Microb Technol 27:347–355.

Tandukar M., Machdar I., Uemura S., Ohashi A., Harada H. (2006) Potential of a combination of UASB and DHS reactor as a novel sewage treatment system for developing countries: long-term evaluation. J Environ Eng 132:166–172.

Templeton M.R., Butler D. (2011) Introduction to Wastewater Treatment (e-book). Bookboon.com. p 80.

Tien M., Kirk T.K. (1988) Lignin peroxidase of *Phanerochaete chrysosporium*. In: Wood W.A., Kellogg S.T. (eds), Methods in Enzymology - Biomass, Part B, lignin, pectin and chitin. San Diego CA, Academic Press. Vol. 161, pp. 238–249.

Triton Electronics (1998) Capillary-Suction-Time Equipment Manual. Triton Electronics Ltd., Dunmow, Essex, England.

Verma A., Shalu, Singh A., Bishnoi N.R., Gupta A. (2013) Biosorption of Cu(II) using free and immobilized biomass of *Penicillium citrinum*. Ecol Eng 61:486–490.

Winkell L., Feldmann J., Meharg A.A. (2010) Quantitative and qualitative trapping of volatile methylated selenium species entrained through nitric acid. Environ Sci Technol 44:382–387.

Wübker S.M., Laurenzis A., Werner U., Friedrich C. (1997) Controlled biomass formation and kinetics of toluene degradation in a bioscrubber and in a reactor with a periodically moved trickle-bed. Biotechnol Bioeng 55:686–692.

Xin B., Xia Y., Zhang Y., Aslam H., Liu C., Chen S. (2012) A feasible method for growing fungal pellets in a column reactor inoculated with mycelium fragments and their application for dye bioaccumulation from aqueous solution. Bioresour Technol 105:100–105.

Xu P., Zeng G.M., Huang D.L., Lai C., Zhao M.H., Wei Z., Li N.J., Huang C., Xie G.X. (2012) Adsorption of Pb(II) by iron oxide nanoparticles immobilized *Phanerochaete chrysosporium*: Equilibrium, kinetic, thermodynamic and mechanisms analysis. Chem Eng J 203:423–431.

Zelmanov G., Semiat R. (2013) Selenium removal from water and its recovery using iron (Fe^{3+}) oxide/hydroxide-based nanoparticles sol (NanoFe) as an adsorbent. Separ Purif Technol 103:167–172.

CHAPTER 5

Sorption of zinc onto elemental selenium nanoparticles immobilized in *Phanerochaete chrysosporium* pellets

A modified version of this chapter has been submitted as:

E.J. Espinosa-Ortiz, M. Shakya, R. Jain, E.R. Rene, E.D. van Hullebusch, P.N.L. Lens (2016). Use of elemental selenium nanoparticles immobilized fungal pellets fungal as sorbent material to remove zinc from water.

Abstract

The use of a novel hybrid biosorbent, elemental selenium nanoparticles (nSe^0) immobilized in pellets of *Phanerochaete chrysosporium,* to remove Zn from aqueous solutions was investigated. Fungal pellets containing nSe^0 (nSe^0-pellets) showed to be better biosorbents as they removed more Zn (88.1 ± 5.3%) compared to Se-free fungal pellets (56.2 ± 2.8%) at pH 4.5 and an initial Zn concentration of 10 mg L^{-1}. The enhanced sorption capacity of nSe^0-pellets was attributed to a higher concentration of sorption sites resulting in a more negative surface charge density, as determined by analysis of the potentiometric titration data. Fourier transform infrared spectroscopy (FT-IR) analysis of fungal pellets prior to and after being loaded with Zn showed the functional groups, including hydroxyl and carboxyl groups, involved in the sorption process. The experimental data indicated that the sorption rate of the nSe^0-pellets fitted well to the pseudo-second order kinetic model (R^2=0.99), and the sorption isotherm was best represented by the Sips model (Langmuir-Freundlich) with heterogeneous factor n=1 (R^2=0.99), which is equivalent to the Langmuir model. Operational advantages of fungal pelleted reactors and the Zn removal efficiencies achieved by nSe^0-pellets under mild acidic conditions make nSe^0-pellet based bioreactors an efficient biosorption process.

Key words: Zinc sorption, fungal pellets, selenium nanoparticles, *Phanerochaete chrysosporium*, hybrid sorbent.

5.1 Introduction

Zinc (Zn) is present in wastewater generated by mineral extraction, metal plating, battery production, tannery processes, pigment and chemical manufacturing, galvanizing plants and textile industries. Zn is a primary contaminant when found at relatively low concentrations, *e.g.* > 3 mg Zn L^{-1} is not permitted in drinking water according to the World Health Organization (2008), due to their toxicological effects for humans.

Conventional physicochemical methods to remove Zn from water and wastewater streams include electrodialysis, electrochemical and chemical precipitation, membrane filtration, ion exchange, coagulation and adsorption (Hua et al., 2012). Most of these methods involve high operating costs and high energy consumption. Biosorption, the adsorption of contaminants from solution by biological material, has emerged as a promising technology due to its simplicity, flexible operational procedures, low cost, high quality effluent and the abundant availability of biomass (Fomina and Gadd, 2014). A broad range of biosorbents for the removal of heavy metals from contaminated water have been demonstrated and reviewed in the literature (Park et al., 2010), including algae, bacteria, yeast, fungi, plant residues and sludge wastes. The use of fungal biomass as a biosorbent material is advantageous, as fungi possess high tolerance towards metals (Valix et al., 2001) and they can grow at low pH conditions, which is characteristic for some Zn containing wastewaters, *e.g.* acid mine drainage and battery making industrial wastewater (Mansoorian et al., 2014). Fungal biomass (dead or active) has been successfully used for the removal of metal ions from aqueous solutions, including Cu^{2+}, Ni^{2+}, Pb^{2+}, Hg^{2+}, $Cr_2O_7^{2-}$, CrO_4^{2-} and Zn^{2+} (Kogej and Pavko, 2001; Bayramoğlu and Arıca, 2008; Filipović-Kovačević et al., 2010).

Engineered nanomaterials such as titanate nanofibers (Xu et al., 2014), metal oxide (Kumar et al., 2013) and magnetic nanoparticles (Ge et al., 2012), as well as biologically produced elemental selenium nanoparticles (nSe^0) (Jain et al., 2015a) efficiently remove a wide range of heavy metals from wastewaters. Particles in the nano-size range possess altered properties compared to their bulk materials, including large surface area and high reactivity (Hua et al., 2012), which makes them attractive as sorbents. However, the main drawbacks of using nanoparticles as sorbents include their reactivity loss due to agglomeration and the difficulty to separate them from the treated effluent (Hua et al., 2012). To overcome these limitations, hybrid sorbents that allow to hold the nanoparticles into a porous supporting material of a large size have been developed (Xu et al., 2012).

The objective of this study was to explore the use of nSe^0 immobilized in *Phanerochaete chrysosporium* pellets (nSe^0-pellets) as a novel hybrid biosorbent to remove Zn from solution. *P. chrysosporium* was chosen due to its heavy metal affinity (Kacar et al., 2002), its ability to grow as pellets and to intracellularly synthesize Se^0 particles in the nano-size range (30-400 nm) upon exposure to selenite (SeO_3^{2-}) (Espinosa-Ortiz et al., 2015a). Batch adsorption experiments were conducted to evaluate the sorption capacity of nSe^0-pellets to remove Zn from aqueous solution. Experimental adsorption data at different pH values, metal concentrations and biomass dose were determined; time dependent studies were carried out as well. The possible mechanisms involved in Zn sorption are discussed based on Fourier transform infrared spectroscopy (FT-IR) spectroscopic studies, potentiometric analysis and modeling of experimental adsorption studies.

5.2 Experimental

5.2.1 Biosorbent preparation

The fungal pellets used as sorbent material were produced as previously described by Espinosa-Ortiz et al. (2015). Briefly, fungal pellets were grown either in the presence or absence of Na_2SeO_3 (10 mg Se L^{-1}) at pH 4.5 and 30 °C in an orbital shaker at 150 rpm. After 4 days of incubation, the fungal pellets were harvested and rinsed with an excess amount of Milli-Q water (18 MΩ-cm) until the pH of the washed solution was near to neutrality. The washed pellets were used for biosorption experiments.

5.2.2 Biosorbent characterization

5.2.2.1 Potentiometric titration

Potentiometric titration experiments were performed with an automatic titrator (Metrohm 848 Tritino Plus, Switzerland). 0.5 g of biomass (Se-free or nSe^0-pellets) was added to a flask containing 30 mL of Milli-Q water and 0.001 M NaCl as background electrolyte concentration. A known amount of 0.1 M NaOH was added at the beginning of the experiment to increase the pH to ~ 8.0 (± 0.3). The suspension was then titrated from pH 8.0 (± 0.3) to 3.0 (± 0.3) using a 29.9 mM HCl solution as the adsorption of Zn was carried out in this range. The change in pH (Δ pH) versus micromoles of HCl added were plotted. The local minima in the graphs represent the minimum change in the pH and hence the pK_a of the functional groups present on the surface of the fungal pellets (Braissant et al., 2007). The acid-base titration data were modeled to determine the surface charge density (C m^{-2}), and functional groups present on the surface of the pellets using the PROTOFIT software (Turner and Fein, 2006) as described by Laurent et al. (2009) and Jain et al. (2015b). Briefly stating, a non-electrostatic adsorption model with four acidic sites and extended Debye-Hückel activity coefficients were used. A four-site model has been used as it is known to predict all the functional groups present on the surface of the biomass as well as describe well the acid-base titration data (Naeem et al., 2006). The model was constrained using the pK_a values determined experimentally from the acid-base titration data. Blank titrations using 0.001 M NaCl solutions were also performed.

5.2.2.2 SEM and FT-IR

Se-free and nSe^0-pellets were morphologically characterized by a JSM-6010 LA analytical scanning electron microscope (SEM) (JEOL, Japan). The samples were fixed in a 2.5% glutaraldehyde solution for 2 h. A series of dehydration steps with 50%, 60%, 70%, 80% and 90% and 100% ethanol were performed (Ngwenya and Whiteley, 2006). The samples were critically point dried, mounted onto a SEM stub and coated with gold. The chemical characteristics of Se-free, nSe^0-pellets and Zn loaded fungal pellets (Se-free and nSe^0-pellets) were analyzed by Fourier transform infrared (FT-IR) spectroscopy (IRAffinity-1, Shimadzu). A mixture of fungal biomass (~0.01 g) with KBr (0.2 g) was prepared and pressed with the help of a bench press. The FT-IR spectrum was then recorded from 4000 to 800 cm^{-1}.

5.2.3 Batch adsorption experiments

The biosorption capacity of nSe^0-pellets was investigated in batch experiments. Fungal biosorbents were mixed with aqueous solutions (10 mL) containing Zn (added as $ZnCl_2$) in 25 mL glass containers and agitated on a rotary shaker at 150 rpm and 30 °C for 24 h. The optimal biomass dose was determined using different biomass concentrations ranging from 0.8-4.0 g dry biomass L^{-1} (10 mg Zn L^{-1}, pH 4.5). The influence of the pH on the Zn adsorption characteristics was studied in a pH range of 2.0-7.0, at a biomass concentration of 2 g L^{-1} and at an initial Zn concentration of 10 mg L^{-1}. The pH was adjusted with 1 M HCl or 1 M NaOH at the beginning of the experiment and their values were not controlled afterwards in the aqueous solution. Biotic control experiments (Se-free pellets) were performed to assess the difference in Zn sorption due to the fungal biomass and the synergetic effect of nSe^0 bearing biomass. Abiotic control experiments (absence of biomass) were done to account for any abiotic factors that might influence the adsorption process. All experiments were done in triplicates.

The Zn removal efficiency of the biosorbents and the metal uptake capacity at equilibrium (q_e) were estimated according to Equations 1 and 2, respectively:

$$Removal\ efficiency, (\%) = \left(\frac{Initial\ concentration\ (mg\ L^{-1}) - Concentration\ at\ t\ (mg\ L^{-1})}{Initial\ concentration\ (mg\ L^{-1})}\right) \times 100$$

Eq. (1)

$$q_e\ (mg\ g^{-1}) = \frac{Volume\ (L) \times [Initial\ concentration\ (mg\ L^{-1}) - Concentration\ at\ t\ (mg\ L^{-1})]}{Dry\ biomass\ (g)}$$

Eq. (2)

Equilibrium sorption isotherm studies were performed at different initial Zn concentrations (10-50 mg Zn L^{-1}, 3 g biomass L^{-1}, pH 4.5). The equilibrium data was analyzed using two-parameter models (Langmuir, Freundlich, Temkin and Dubinin-Radushkevich) and a three-parameter model (Sips, which is a combination of the Langmuir-Freundlich models), as described elsewhere (Table 5.1). The four types of linearized Langmuir equations were used (Bolster and Hornberger, 2007). The q_e values obtained from the experimental data (Equation 2) were used in the isotherm models to describe the sorption behavior and to calculate the maximum metal uptake capacity (q_m). Sorption kinetics were determined by carrying out a time-dependency study at different contact times (15-720 min, 10 mg Zn L^{-1}, 3 g biomass L^{-1}, initial pH 4.5). The data was fitted to the pseudo-first and pseudo-second order kinetic models (Table 5.1).

5.2.4 Analytical methods

After the sorption experiments, solid-liquid separation was carried out by centrifuging at 37,000 g for 15 min at 4 °C (Hermle Z36 HK, HERMLE Labortechnik GmbH, Germany). The biomass concentration was determined gravimetrically as dry weight by filtering the biomass suspension through a pre-dried (24 h at 105 °C) and pre-weighted 0.45 μm filter paper (type GF/F, Whatman Inc., Florham Park, NJ). The supernatant was used to determine the residual total Zn concentration in the sample measured by Atomic Absorption Spectroscopy (AAS200, Perkin Elmer) at 213.9 nm. To corroborate that selenium was not released from the nSe^0-pellets to the aqueous phase, the total selenium concentration was measured in the supernatant as described previously in the literature (Espinosa-Ortiz et al., 2015a).

Table 5.1 Isothermic and kinetic models used in this study.

Isotherm model	Equation	Linearization	References
Langmuir	$q_e = \frac{q_m k_L C_e}{1 + k_L C_e}$	Type I (Hanes-Woolf) $\frac{C_e}{q_e} = \frac{1}{q_m k_L} + \frac{C_e}{q_m}$ Type II (Burke) $\frac{1}{q_e} = \frac{1}{q_m K_L C_e} + \frac{1}{q_m}$ Type III (Eadie-Hofstee) $q_e = q_m - \frac{1}{K_L}\frac{q_e}{C_e}$ Type IV (Scatchard) $\frac{q_e}{C_e} = K_L q_m - K_L q_e$	Foo and Hameed, 2010; Worch, 2012; Bolster and Hornberger, 2007
Freundlich	$q_e = k_f C_e^{1/n}$	$Log(q_e) = \frac{1}{n} log(C_e) + \log(k_f)$	
Temkin	$q_e = B \ln(A_T C_e)$ $B = \frac{RT}{b}$	$q_e = B \ln A_T + B \ln C_e$	
Dubini-Radushkevich	$q_e = q_m e^{(-k\epsilon^2)}$ $\epsilon = RT \ln\left(1 + \frac{1}{C_e}\right)$	$\ln q_e = \ln q_m - k\, \epsilon^2$	
Sips (Freundlich-Langmuir)	$q_e = \frac{C_e^{1/n}}{\left(1/q_m k_s\right)\left(C_e^{1/n}/q_m\right)}$		
Kinetic model	**Equation**	**Linearization**	**References**
Pseudo-first order	$q_t = q_e[1 - exp(-k_1 t)]$	$\log(q_e - q_t) = \log q_e - \left[\frac{k_1}{2.303} t\right]$	Lin and Wang, 2009
Pseudo-second order	$q_t = \frac{k_2 q_e^2 t}{1 + k_2 q_e t}$	$\frac{t}{q_t} = \frac{1}{k_2 q_e^2} + \left[\frac{1}{q_e} t\right]$	

5.3 Results

5.3.1 Characterization of biosorbent material

Se-free and nSe^0-pellets showed a different macro-morphology: Se-free pellets were hairy and fluffy with a white-ivory color (Figure 5.1A), whereas nSe^0-pellets were smooth and compact with a red-orange coloration (Figure 5.1B), indicating the presence of Se^0. Differences in the micro-morphology were also observed: Se-free pellets showed a fibril-like structure, presenting a porous surface with several channels (Figure 5.1C), whereas nSe^0-pellets showed a smooth surface with smaller openings between the hyphae (Figure 5.1D). Se^0 synthesized by *P. chrysosporium* are spherical particles in the nano-size range (30-400 nm) (Espinosa-Ortiz et al., 2015a).

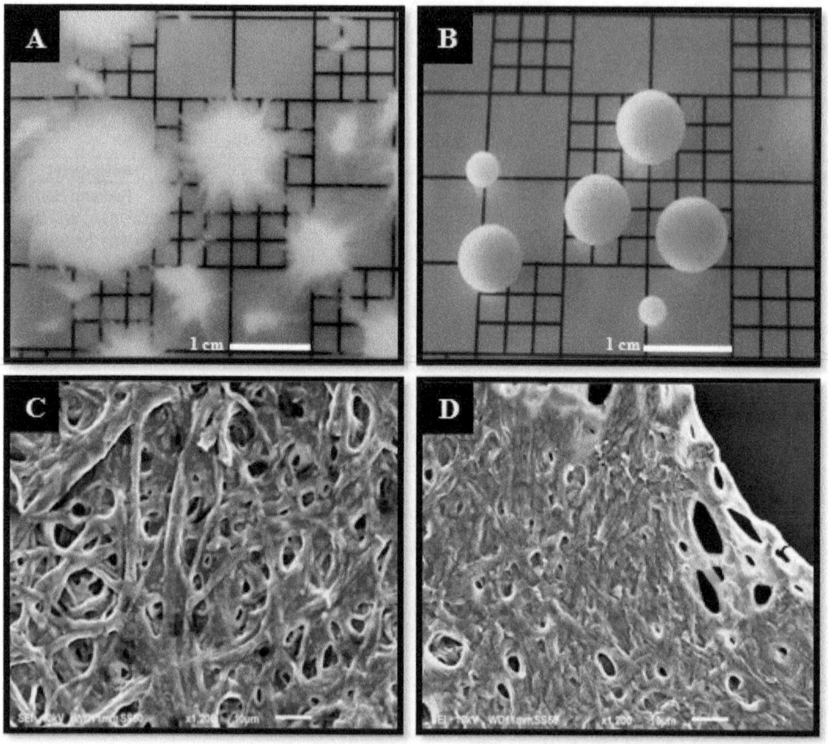

Figure 5.1 Characterization of fungal biosorbents. Macroscopic image of A) Se-free pellets and B) nSe^0-pellets. C) Scanning electron micrograph of C) Se-free pellets and D) nSe^0-pellets.

The acid-base titration was able to predict three local minima of Se-free and nSe^0-pellets (Figure 5.2A), corresponding to the pK_a values of the major functional groups. As the change in the pH is very small at the beginning and end of titration, the acid-base titration experiments may miss the functional groups whose pK_a are at the extreme points of the acid-base titration (Braissant et al., 2007). However, the modeling of the acid-base titration data was not only able to identify all the functional groups, including those missed by the experimental data, but it was also able to estimate the concentration of the functional groups, as also observed during the acid-base titration of selenium fed activated sludge (Jain et al., 2015). Table 5.2 summarizes the experimental and calculated pK_a values, the corresponding site concentrations and assignation of the functional groups. The simulated titration obtained by the non-electrostatic model fitted well with the experimental data (Figure 5.2B). The surface charge density of the nSe^0-pellets was between 1.9 and 120 times more negative than the Se-free pellets in a pH range between 7.5 and 3.5, respectively (Figure 5.2C).

Table 5.2 Estimated proton binding constants (pKa), functional groups and active site concentrations for Se-free and nSe^0-pellets.

Biosorbent	pK$_a$	pK$_a$ modeled	Site concentration (mol g^{-1})	Functional groups	References
Se-free pellets	3.8	3.8	5.6 x 10^{-8}	Phosphodiester/carboxyl	Luo et al.,2014; Tourney et al., 2014; Baker et al., 2010; Deng et al., 2005
	7.2	7.24	5.0 x 10^{-6}	Phosphoryl	
	7.6-8.1	7.82	2.0 x 10^{-5}	Sulfonic/sulfinic/thiols	
	ND	10.3	3.2 x 10^{-8}	Hydroxyl/amine	
nSe^0-pellets	3.5	3.5	4.7 x 10^{-6}	Phosphodiester/carboxyl	
	4.4	4.4	1.5 x 10^{-6}	Carboxyl	
	6.8	6.88	1.6 x 10^{-5}	Phosphoryl	
	ND	10.0	2.4x 10^{-6}	Hydroxyl/amine	

Note: ND=Not determined

The FT-IR spectra confirmed the presence of ten absorption bands for Se-free and nSe^0-pellets prior to and after exposure to Zn solution (Figure 5.3). The broadest IR absorption, around 3400 cm^{-1}, corresponds to the overlapping of -OH and -NH_2 peaks (hydroxyl and amino groups) (Kumar and Min, 2011). The bands observed at around 2920 cm^{-1} and 2855 cm^{-1} represent aliphatic groups (-CH) with both asymmetric and symmetric stretching (Dong et al., 2013; Lecellier et al., 2014). The bands at 1720-1740 cm^{-1} indicate the stretching of C=O, which may originate from carboxylic groups (esters and fatty acids) (Dong et al., 2013, Lecellier et al., 2014). The 1653 cm^{-1} and 1559 cm^{-1} bands correspond to the amide I (C=O stretching) and amide II (C-N stretching) arising from the presence of proteins, respectively (Bai and Abraham, 2002). The 1450 cm^{-1} band indicates -CH_2/CH_3 bending (lipids) (Marshall et al., 2005). The band at 1372 cm^{-1} was attributed to the bending of -CH_3. The 1154 cm^{-1} band is related to the stretching of P=O of the phosphate group (Arica and Bayramoğlu, 2007). The C-O stretching band of the carbohydrates was observed at 1070 cm^{-1}.

5.3.2 Effects of operational parameters on sorption capacity of nSe^0-pellets

5.3.2.1 Effect of pH on Zn sorption

Figure 5.4A shows the biosorption capacity of *P. chrysosporium* for Zn (10 mg Zn L^{-1}) at different pH values (2.0-7.0). The equilibrium pH varied from 1.7 to 3.5 and 1.6 to 6.3 for Se-free and nSe^0-pellets, respectively. The removal of Zn by *P. chrysosporium* pellets (with and without Se) was found to be strongly pH dependent, *i.e.*, the sorption capacity increased as the pH values increased. The maximum Zn removal efficiencies were observed at pH values between 5.0 and 7.0, with average removal efficiencies of 48.6 (± 4.5)% and 76.7 (± 1.7)% for Se-free and nSe^0-pellets, respectively. At initial pH values < 3.0, uptake of Zn was almost negligible (<5%) for both Se-free and nSe^0-pellets. The removal efficiency of Se-free pellets did not reach 50%, whereas the removal efficiency of nSe^0-pellets exceeded 50% when the initial pH was between 3.0 and 4.0. The simulations by Visual MINTEQ confirmed the absence of Zn precipitation at the equilibrium concentrations tested in this study (data not shown).

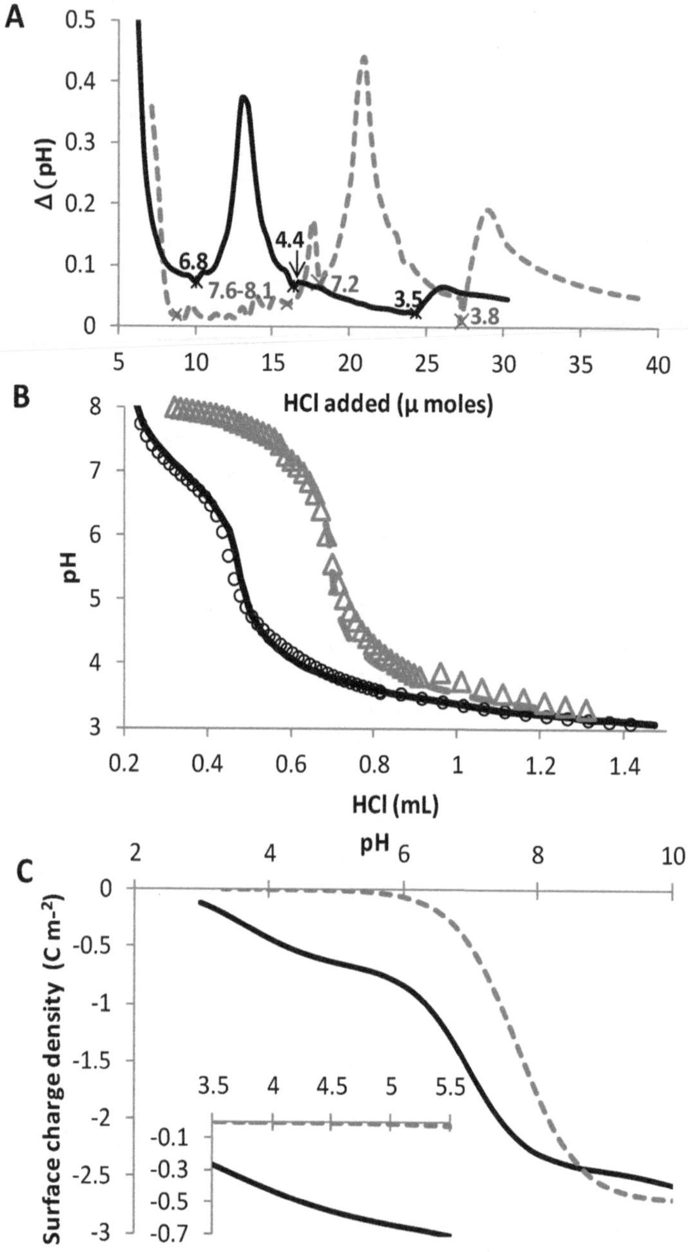

Figure 5.2 Experimental pKa values (Δ pH versus μ moles HCl added) for Se-free (‒ ‒) and nSe^0-pellets (▬▬), B) Titration data for Se-free (△ Raw, ‒ ‒ Simulated) and nSe^0-pellets (○ Raw, ▬▬ Simulated), and C) Surface charge density for Se-free (‒ ‒) and nSe^0-pellets (▬▬). Note: 29.9 mM HCl were used to titrate the fungal suspension (0.5 g biomass, 30 mL MQ-water, 0.001 M NaCl).

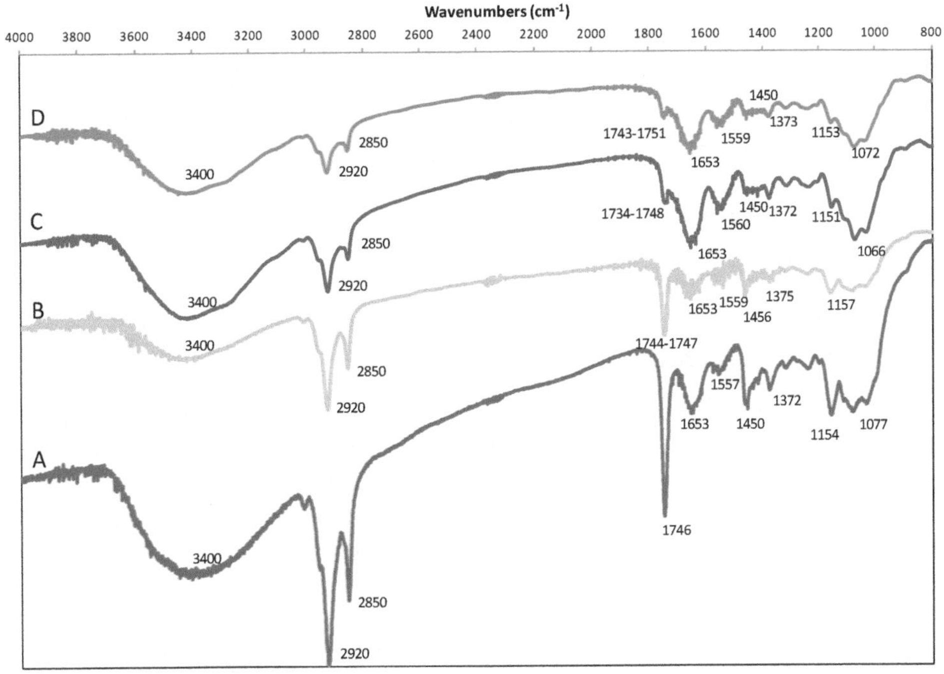

Figure 5.3 FT-IR spectra of fungal pellets. A) Se-free, B) Se-free + Zn, C) nSe^0-pellets, D) nSe^0-pellets + Zn.

5.3.2.2 Effect of biosorbent dose on Zn sorption

An increase of biosorbent dose from 0.8 to 3.2 g L^{-1} resulted in higher removal of Zn, from 45% to 85% in the case of nSe^0-pellets and from 40% to 70% in Se-free pellets (Figure 5.4B). However, an increase of biosorbent dose above 3.2 g L^{-1} did not increase the metal removal efficiency further. Therefore, 3.2 g L^{-1} was chosen as the optimal biosorbent dose for further sorption experiments, with an initial pH value of 4.5 and equilibrium pH of 3.8 and 5.2 for Se-free and nSe^0-pellets, respectively.

5.3.3 Sorption kinetics

The kinetic profiles of Zn biosorption by Se-free and nSe^0-pellets at 10 mg Zn L^{-1} and an equilibrium pH of 3.4 and 5.4, respectively, in Figure 5.4C. The Zn removal rate was fast during the initial stages of the biosorption process (0-60 min, see inset Figure 5.4C). An increase in the Zn removal efficiency over time was observed. The equilibrium value was reached at about 240 min for nSe^0-pellets and at about 480 min for Se-free pellets, with a maximum removal efficiency of 89% and 49%, respectively. The experimental data was modeled using pseudo-first and pseudo-second order reactions (See Appendix, Figures A1 and A2). The rate law for a pseudo-second order kinetic model best described the experimental data (R^2=0.99) for both Se-free and nSe^0-pellets (Table 2). The equilibrium concentrations calculated (q_e) from this model for both Se-free (1.79 mg g^{-1}) and nSe^0-pellets (3.03 mg g^{-1}) were in accordance with the experimental values (Se-free, 1.73 mg g^{-1}; nSe^0-pellets, 3.07 mg g^{-1}).

Table 5.3 Sorption data evaluated by the pseudo-first and pseudo-second order kinetic models.

Biosorbent	Pseudo-first order			Pseudo-second order		
	$q_{e\ cal}$ (mg g^{-1})	k_1 (min^{-1})	R^2	$q_{e\ cal}$ (mg g^{-1})	k_2 (g mg^{-1} min^{-1})	R^2
Se-free pellets	0.80	0.01	0.74	1.79	0.02	0.99
nSe^0-pellets	1.33	0.03	0.89	3.03	0.1	0.99

5.3.4 Adsorption isotherms

The Zn biosorption capacity of Se-free (Figure 5.4D) and nSe^0-pellets (Figure 5.4E) as a function of the initial Zn concentration (10-50 mg Zn L^{-1}) in aqueous solution was determined. The equilibrium sorption capacity increased with increasing initial Zn concentration, from 1.9 to 8.3 mg g^{-1} for Se-free pellets and from 2.8 to 11.3 mg g^{-1} in nSe^0-pellets. A plateau was reached at an initial Zn concentration of 40 mg L^{-1}. The removal efficiency decreased with increasing initial Zn concentration, from 82% to 38% in nSe^0-pellets, and from 60% to 30% in Se-free pellets.

The adsorption isotherms were fitted to the Langmuir, Freundlich, Temkin, Rubinin- Radushkevich and Sips models (Table 5.3). All Langmuir linearization types fitted the data for nSe^0-pellets ($R^2 \geq 0.98$) the best (See Appendix, Figure A3). Linearization II (Burke's model) had the highest R^2 value (0.99), with a maximum sorption capacity (q_m) of 13.9 mg g^{-1} and an affinity constant (k_L) of 0.15 L mg^{-1}. None of the linearized forms of Langmuir models fitted the experimental data for the data from Se-free pellets ($R^2<0.97$) (See Appendix 1, Figure A3), with overestimated q_m values (up to 40 mg g^{-1}) and low k_L (≤ 0.06 L mg^{-1}), suggesting a lower sorption affinity and a less favorable sorption towards Zn compared to nSe^0-pellets. The Freundlich, Temkin and Dubinin-Radushkevich models did not represent well the experimental data of both biosorbents tested with R^2 values ≤ 0.97 (See Appendix 1, Figure A4). With a heterogeneous factor (n) of 1 the Sips equation reduces to Langmuir, which showed the best fit for the nSe^0-pellets as described above. The Sips model was the best fit for the Se-free pellets with n, q_m and k_s values of 0.6, 10.2 mg g^{-1} and 0.02 L mg^{-1} (R^2=0.99), respectively, suggesting a more heterogeneous system than the nSe^0-pellets. The Sips model with n=1, equivalent to the Langmuir isotherm model, fitted well for nSe^0-pellets, indicating that Zn sorption takes place at finite and specific homogeneous binding sites, most likely acidic sites with pK_a values between 3.5 and 4.4 (Table 5.2), which are energetically equivalent. This suggests that the organization of the sorbed Zn occurs in a monolayer with no interaction among them.

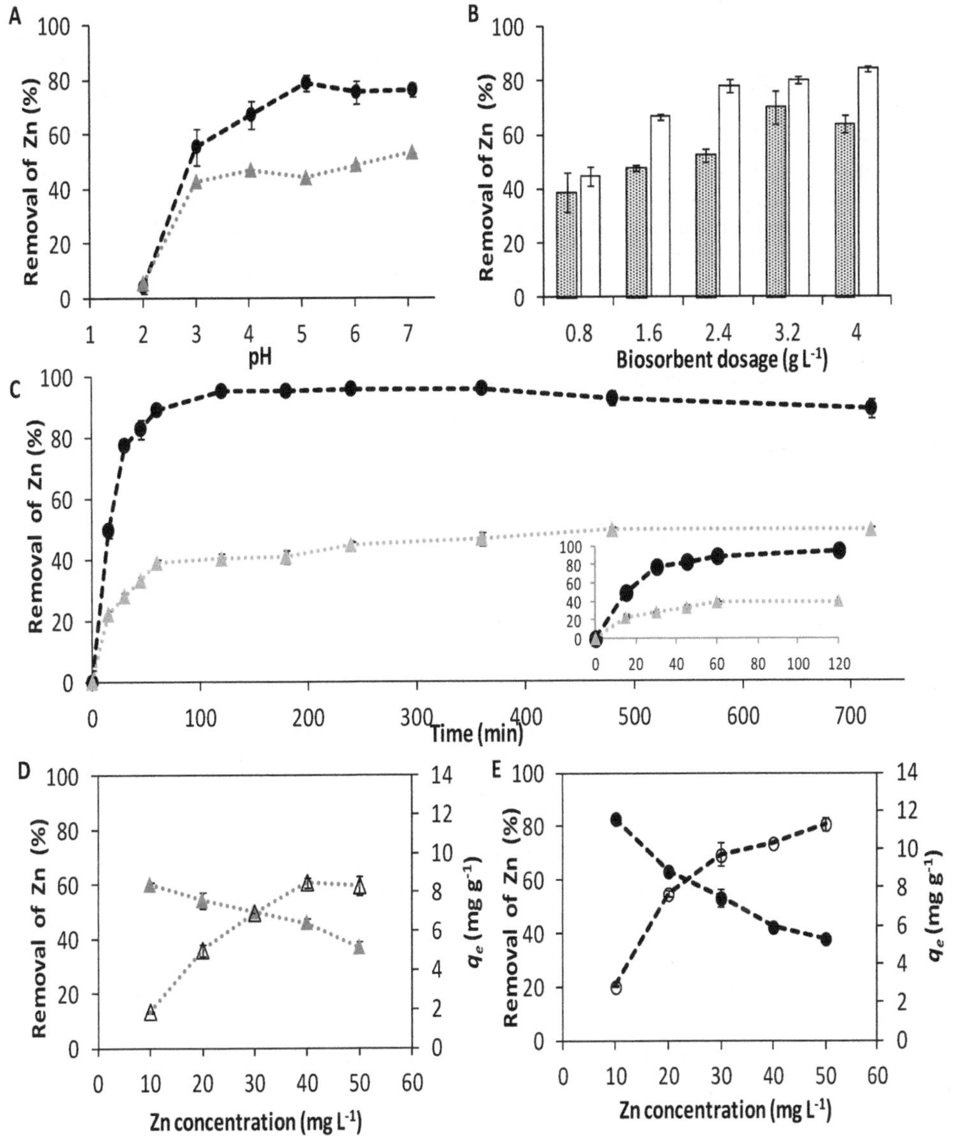

Figure 5.4 Effects of operational parameters on the biosorption capacity of Zn on Se-free (▲) and nSe⁰-pellets (●—). A) pH, B) Biomass dosage (▨ Se-free and ☐ nSe⁰-pellets), C) Time dependency, and adsorption isotherms of D) Se-free (△ Zn removal, △ q_e) and E) nSe⁰-pellets (● Zn removal, ○ q_e). Biosorption conditions: Initial Zn concentration, 10 mg L⁻¹; volume of biosorption medium, 10 mL; temperature, 30 °C; biosorption time, 24 h; initial pH, 4.5.

Table 5.4 Isotherm constants for Zn adsorption onto Se-free and nSe^0-pellets.

Biosorbent	Langmuir model											
	Type I (Hanes-Woolf)			Type II (Burke)			Type III (Eadie-Hofstee)			Type IV (Scatchard)		
	q_m	k_L	R^2	q_m	k_L	R^2	q_m	k_L	R^2	q_m	k_L	R^2
Se-free pellets	15.9	0.04	0.80	40.1	0.01	0.97	13.2	0.06	0.40	23.9	0.02	0.40
nSe^0-pellets	13.3	0.17	0.99	13.9	0.15	0.99	13.2	0.16	0.98	13.7	0.15	0.98

	Freundlich			Temkin			Dubinin-Radushkevich			Sips (Langmuir-Freundlich)			
	k_F	n	R^2	A_T	B	R^2	q_m	k_{DB}	R^2	q_m	k_S	n	R^2
Se-free pellets	0.79	1.33	0.91	0.47	3.37	0.96	8.1	5×10^{-6}	0.97	10.2	0.02	0.6	0.99
nSe^0-pellets	2.47	2.09	0.94	0.46	0.80	0.96	10.0	9×10^{-7}	0.96	13.9	0.15	1	0.99

Note: q_m (mg g^{-1}), k_L (L mg^{-1}), k_F (mg g^{-1})(L mg^{-1})$^{1/n}$, A_T (L mg^{-1}), B (J mol^{-1}), k_{DB} (mol^2 (kJ^2)$^{-1}$), k_S (L mg^{-1})

5.4 Discussion

5.4.1 Sorption mechanisms of Zn onto nSe^0 pellets

This study showed that nSe^0-pellets have a higher sorption capacity towards Zn compared to Se-free pellets of *P. chrysosporium*. Se-free and nSe^0-pellets contained similar functional groups, with differences in the surface charge density and number of active sites (Table 5.2). nSe^0-pellets presented a more negative surface charge density as compared to the Se-free pellets (1.9 times higher at pH 7.5 and 120 times higher at pH 3.5) (Figure 5.2C), which could explain a better performance of nSe^0-pellets as sorbent material for Zn. This was further evident by the larger site density of phosphodiester and carboxyl groups for nSe^0-pellets when compared to Se-free pellets (Table 5.1), which would be largely deprotonated at the pH values used in this study. Furthermore, the pK_a of the phosphodiester and carboxyl groups was lower than that of the Se-free pellets, thus providing more deprotonated sites for Zn adsorption. A higher concentration of active sites for the carboxylic group could also explain a higher adsorption capacity of the nSe^0-pellets.

Determination of the FT-IR spectra of the fungal biomass prior to and after Zn loading (Figure5.3) provided an idea of the functional groups involved in the sorption process. The involvement of carboxylic groups in the sorption of Zn could be indicated by the slight shift in the bands observed from 1746 to 1744-1747, and from 1734-1748 to 1743-1751, for Se-free and nSe^0-pellets, respectively. Furthermore, both Se-free and nSe^0-pellets have high concentrations of carboxylic groups (Table 5.2), which carry negative charges and play a key role as scavengers of metal ions in solution as observed during the sorption of Pb^{2+} when using *R. arrhizus* biomass (Naja et al., 2005). The spectrum of Se-free pellets at 1077 cm^{-1} (carbohydrate backbone) disappeared after Zn biosorption, and shifted from 1072 cm^{-1} to 1066 cm^{-1} in the case of nSe^0-pellets.

The Sips model with n=1, equivalent to the Langmuir isotherm model, fitted well for nSe^0-pellets, indicating that Zn sorption takes place at finite and specific homogeneous binding sites, most likely acidic sites with pK_a values between 3.5 and 4.4 (Table 5.2), which are energetically equivalent. This suggests that the organization of the sorbed Zn occurs in a monolayer with no interaction among them. The adsorption of heavy metals (*e.g.* Cu, Pb, Zn, Ni) by other fungal species has also shown to fit well to Langmuir isotherms (Bayramoğlu et al., 2003; Javaid et al., 2011). The Se-free pellets were better represented by the Sips equation with a heterogeneity factor of 0.6, indicating a more heterogeneous Zn sorption process compared to that of the nSe^0-pellets. Table 5.4 compares the maximal sorption capacities obtained in this study with values of other sorbents removing Zn from acidic effluents. Treatment of acidic effluents polluted with heavy metals is challenging due to the competition for the sorption sites between metal ions and protons in solution, decreasing the adsorption capacity of the sorbent materials. Among the different sorbents tested and reported in the literature (Table 5.4), the use of nSe^0-pellets for the removal of Zn under mild acidic conditions (pH 4.5) is promising, showing similar or better performance than other biosorbents.

From the kinetic viewpoint, the sorption of Zn onto Se-free and nSe^0-pellets occurred in a two-step process: i) a fast sorption of Zn to the outer part of the biosorbent (≤ 60 min); and ii) a slow intracellular diffusion in the interior of the biosorbent (Figure 5.4C). The experimental data of both Se-free and nSe^0-pellets was best fitted to the pseudo-second order kinetic model, suggesting that the sorption mechanism mutually depended on both the Zn solution and the type of biosorbent, and that the rate-limiting step might be due to chemical sorption (Ho and McKay, 1999). The pseudo-second order sorption rate constant values obtained in this study were low ($k_2 < 0.05$), which indicates that the sorption rate is proportional to the number of available bonding sites, which decreases with an increase in the contact time. The removal of Pb by a hybrid sorbent composed of iron oxide magnetic nanoparticles and Ca-alginate immobilized *P. chrysosporium* also occurred by a two-step sorption process (Xu et al., 2012), a rapid sorption to the surface and a slow intra-particle diffusion. According to the authors, the sorption of Pb to this hybrid sorbent was suggested to be a combination of the biosorption affinity of *P. chrysosporium* and its enhanced intracellular accumulation by the iron oxide nanoparticles as indicated by the observation of Fe-O groups which may form chemical bonds with Pb (Xu et al., 2012).

An active role of nSe^0 in the sorption of Zn by nSe^0-pellets of *P. chrysosporium* is not evident from the results obtained in this study as the participation of similar functional groups in Zn sorption was observed in both Se-free and nSe^0-pellets. However, it is clear that the exposure of *P. chrysosporium* pellets to SeO_3^{2-} leads to the modification of the fungal biomass and its sorption characteristics. It has been suggested that the amount and composition of extracellular polymeric substances can be modified by the presence of SeO_3^{2-}, which can in turn be one of the explanations for the difference in the surface charge density of the nSe^0-pellets (Jain et al., 2015; Xu et al., 2009).

5.4.2 Effects of operational parameters on sorption capacity of nSe^0-pellets

The solution pH drives the biosorption process by determining the speciation of metals in solution. Anew, pH also influences the surface properties of the sorbents, affecting the dissociation of binding sites and surface charge (Pagnanelli et al., 2003), which varies according to the type of biosorbent and the metal ion to be removed. In this study, the Zn sorption capacity of Se-free and nSe^0-pellets was pH dependent, increasing as the solution pH values increased. A low adsorption capacity is typically observed at pH values below 4.0, which can be attributed to the competition effects of Zn^{2+} with H^+ ions. Dissociation of protons from the carboxylic acid group, one of the major components of the fungal cell wall (Zaidi et al., 2011), increases as the pH of the aqueous solution increases, thus increasing the negative groups in the fungal surface and therefore the number of binding sites available for complexation of metal cations (Tobin et al., 1994).

Table 5.4 Comparison of the maximum Zn sorption capacity of nSe^0-pellets with other sorbents.

Adsorbent type	Description	pH	T (°C)	q_m (mg g^{-1})	Reference
Algae	*Ulva fasciata*	5.0	30	13.5	Kumar et al., 2006
Bacteria	*Pseudomonas putida*	5.0	30	27.4 (living) 17.7 (non-living)	Chen et al., 2005
Yeast	*Candida utilis* (Ca-alginated immobilized)	5.17	25	149.7	Ahmad et al., 2013
	Candida tropicalis (Ca-alginated immobilized)	5.17	25	119.9	Ahmad et al., 2013
Fungi	*P. ostreatus*	5.0	25	3.22	Javaid et al., 2011
	P. chrysosporium	5.0	--	7.75	Yan and Viraraghavan, 2003
	nSe^0- pellets of *P. chrysosporium*	4.5	30	13.9	Present study
Chemical	Activated carbon	4.5	25	31.1	Mohan and Singh, 2002
Nanoparticles	nSe^0 biologically produced	6.5	30	60	Jain et al., 2015

Note: *q_m=maximum sorption capacity

In general, fungal surfaces are negatively charged in the pH range of 2.0 to 6.0 (Figure 5.2). Biosorption of Pb by *Aspergillus parasiticus* was strongly reduced when the pH of the solution was below 3.0 (Akar et al., 2007). Similarly, the removal of Hg, Cd and Zn by *Funalia trogii* immobilized in Ca-alginate increased with an increase in the pH from 3.0 to 6.0, and optimal removal efficiencies were obtained at pH 6.0 (Arica et al., 2003). It is interesting to note that the after the adsorption of Zn onto Se-free and nSe^0-pellets, the equilibrium pH decreased and increased, respectively, for Se-free and nSe^0-pellets adsorption system. The increase in equilibrium pH was also observed in the case of Zn adsorption onto Bio Se nanoparticles (Jain et al., 2015). The increase in the equilibrium pH is suggested due to the adsorption of H^+ or release of OH^- ion suggesting the ligand-like type II reaction (Jain et al., 2014) which usually takes place at lower initial pH as used in this study. The decrease in the equilibrium pH during the Zn adsorption on Se-free pellets is most likely due to replacement of H^+ by Zn ions.

An increased amount of fungal biomass resulted in a better sorption of Zn for Se-free and nSe^0-pellets. This might be associated to the higher availability of surface area and therefore more binding sites with increased biomass dose. Increasing the biosorbent concentration beyond the optimum biosorbent dose (3.2 g L^{-1}) did not further increase the sorbent capacity, as previously observed (Parvathi et al., 2007). The decreased adsorption capacity at increasing biomass dose can be attributed to different factors including the availability of the solute and the decrease in the surface area of the sorbent by the formation of biomass aggregates (Marandi et al., 2010).

5.4.3 Potential applications

This study showed the ability of nSe^0-pellets as sorbent material for the removal of Zn. This hybrid biosorbent performed well in the pH range of 3.0-7.0. nSe^0-pellets were able to remove Zn under mildly acidic conditions (3.0-4.0), with Zn removal efficiencies exceeding 50%. This is promising for the treatment of heavy metals in (mild) acidic effluents, including acid mine drainage. Other sorbents as bacteria, yeasts and activated carbon have a higher sorption capacity than nSe^0-pellets (Table 5.4) in a similar pH range (4.5-5.0). Bacterial produced Se^0 nanoparticles showed a higher adsorption capacity towards Zn (60 mg g^{-1}), although this value was obtained at a higher pH (6.5) (Jain et al., 2015). However, compared to other biosorbents, including algae and fungal biomass (chemically pre-treated or not), nSe^0-pellets showed a higher sorption capacity for Zn.

Though some sorbents showed better sorption capacities than nSe^0-pellets, there are some operational advantages of using this hybrid sorbent. Entrapping the nSe^0 in the fungal biomass avoids the washout of the nanoparticles along with the effluent. Moreover, the use of fungal pellets facilitates the separation of the biosorbent from the effluent, offering the possibility to easily retain the Zn loaded biomass. The use of *P. chrysosporium* pellets to continuously remove SeO_3^{2-} and produce nSe^0 has already been shown in an up-flow reactor (Espinosa-Ortiz et al., 2015b). This paper expands the potential of such a continuous system with a biosorption feature, which facilitates the simultaneous removal of selenium and heavy metals from polluted effluents.

Further research is required to evaluate the performance of the nSe^0-pellets in multi-metal solution systems, to determine if the hybrid fungal sorbent can also bind metals selectively as suggested for bacterial Se^0 nanoparticles (Jain et al., 2016). Moreover, the regeneration capacity of the nSe^0-pellets and the recovery of Zn from the concentrated leaching solution should be further investigated.

5.4.4 Conclusions

This study demonstrates the use of fungal nSe^0-pellets as a novel material for the sorption of Zn from wastewater in the pH range of 3.0-7.0. The biosorption capacity of the fungal pellets was enhanced when nSe^0 were entrapped in the biomass. Operational parameters such as pH, initial Zn concentration and biomass dose influenced the Zn removal efficiency. The equilibrium sorption data were well represented by the pseudo-second order kinetic model, suggesting a two-step process including a fast sorption of Zn to the surface ($\leq$ 60 min) and a slow intracellular diffusion inside the fungal pellets. The higher concentration of carboxyl and phosphodiester groups on the nSe^0-pellets compared to Se-free pellets resulted in a more negative surface charge on nSe^0 pellets (pH: 3.0-7.5), yielding better biosorbent performance. The insights gained from this study and the operational advantages of fungal pelleted reactors suggest the application of this bioprocess to simultaneously remove SeO_3^{2-} and Zn from wastewaters.

5.5 References

Bai R.S., Abraham T.E. (2002) Studies on enhancement of Cr (VI) biosorption by chemically modified biomass of *Rhizopus nigricans*. Wat Resear 36:1224–1236.

Bayramoğlu G., Arıca M.Y. (2008) Removal of heavy mercury(II), cadmium(II) and zinc(II) metal ions by live and heat inactivated *Lentinus edodes* pellets. Chem Eng J 143:133–140.

Braissant O., Decho A.W., Dupraz C., Glunk C., Przekop K.M., Visscher P.T. (2007) Exopolymeric substances of sulfate-reducing bacteria: Interactions with calcium at alkaline pH and implication for formation of carbonate minerals. Geobiology 5(4):401–411.

Chen X.C., Wang Y.P., Lin Q., Shi J.Y., Wu w.X., Chen Y.X. (2005) Biosorption of copper (II) and zinc (II) from aqueous solution by Pseudomonas putida CZ1. Colloid Surf B: Biointrfaces 46(2):101–107.

Dong X.Q., Yang J.S., Zhu N., Wang E.T., Yuan H.L. (2013) Sugarcane bagasse degradation and characterization of three white-rot fungi. Bioresour Technol 131:443–451.

Espinosa-Ortiz E.J., Gonzalez-Gil G., Saikaly P.E., van Hullebusch E.D., Lens P.N.L. (2015a) Effects of selenium oxyanions on the white-rot fungus *Phanerochaete chrysosporium*. Appl Microbiol Biotechnol 99(5):2405-2418.

Espinosa-Ortiz E.J., Rene E.R., van Hullebusch E.D., Lens P.N.L. (2015b) Removal of selenite from wastewater in a *Phanerochaete chrysosporium* pellet based fungal bioreactor. Int Biodeterior Biodegradation 102:361–369.

Filipović-Kovačević Z., Sipos L., Briški F. (2010) Biosorption of chromium, copper, nickel, and zinc ions onto fungal pellets of *Aspergillus niger* 405 from aqueous solutions. Food Technol Biotechnol 38:211–216.

Fomina M., Gadd G.M. (2014) Biosorption: current perspectives on concept, definition and application. Bioresour Technol 160:3–14.

Ge F., Li M.M., Ye H., Zhao B.X. (2012) Effective removal of heavy metal ions Cd^{2+}, Zn^{2+}, Pb^{2+}, Cu^{2+} from aqueous solution by polymer-modified magnetic nanoparticles. J Hazard Mater 211–212:366–72.

Ho Y.S., Mckay G. (1998) The kinetics of sorption of basic dyes from aqueous solution by sphagnum moss peat. Canadian J Chem Eng 76(4):822–827.

Hua M., Zhang S., Pan B., Zhang W., Lv L., Zhang Q. (2012) Heavy metal removal from water/wastewater by nanosized metal oxides: a review. J Hazard Mater 211-212:317–31.

Jain R., Jordan N., Schild D., van Hullebusch E.D., Weiss S., Franzen C., Farges F., Hübner R., Lens P.N.L. (2015a) Adsorption of zinc by biogenic elemental selenium nanoparticles. Chem Eng J 260:855–863.

Jain R., Seder-Colomina M., Jordan N., Dessi P., Cosmidis J., van Hullebusch E.D., Weiss S., Farges F., Lens P.N.L. (2015b) Entrapped elemental selenium nanoparticles affect physicochemical properties of selenium fed activated sludge. J Hazard Mater 295:193–200.

Javaid A., Bajwa R., Shafique U., Anwar J. (2011) Removal of heavy metals by adsorption on Pleurotus ostreatus. Biomass Bioener 35(5):1675–1682.

Kacar Y., Arpa C., Tan S., Denizli A., Genc O., Yakup Arica M. (2002) Biosorption of Hg (II) and Cd (II) from aqueous solutions : comparison of biosorptive capacity of alginate and immobilized live and heat inactivated *Phanerochaete chrysosporium*. Process Biochem 37:601–610.

Kogej A., Pavko A. (2001) Laboratory experiments of lead biosorption by self-immobilized *Rhizopus nigricans* pellets in the batch stirred tank reactor and the packed bed column. Chem Biochem Eng 15:75–79.

Kumar K.Y., Muralidhara H.B., Nayaka Y.A., Balasubramanyam J., Hanumanthappa H. (2013) Low-cost synthesis of metal oxide nanoparticles and their application in adsorption of commercial dye and heavy metal ion in aqueous solution. Powder Technol 246:125–136.

Kumar N.S., Min K. (2011) Phenolic compounds biosorption onto *Schizophyllum commune* fungus: FTIR analysis, kinetics and adsorption isotherms modeling. Chem Eng J 168:562–571.

Kumar Y.P., King P., Prasad V.S.R.K. (2006) Comparison for adsorption modelling of copper and zinc from aqueous solution by *Ulva fasciata* sp. J Hazard Mat 137:1246–1251.

Laurent J., Casellas M., Dagot C. (2009) Heavy metals uptake by sonicated activated sludge: relation with floc surface properties. J Hazard Mater 162:652–660.

Lecellier A., Mounier J., Gaydou V., Castrec L., Barbier G., Ablain W., Manfait M., Toubas D., Sockalingum G.D. (2014) Differentiation and identification of filamentous fungi by high-throughput FTIR spectroscopic analysis of mycelia. Int J Food Microbiol 168-169:32–41.

Lin J., Wang L. (2009) Comparison between linear and non-linear forms of pseudo-first-order and pseudo-second-order adsorption kinetic models for the removal of methylene blue by activated carbon. Front Environ Sci Eng 3(3):320–324.

Mansoorian H.J., Mahvi A.H., Jafari A.J. (2014) Removal of lead and zinc from battery industry wastewater using electrocoagulation process: Influence of direct and alternating current by using iron and stainless steel rod electrodes. Sep Purif Technol 135:165–175.

Naja G., Mustin C., Volesky B., Berthelin J. (2005) A high-resolution titrator: A new approach to studying binding sites of microbial biosorbents. Water Res 39:579–588.

World Health Organization (2008) Guidelines for Drinking-water Quality.

Pagnanelli F., Esposito A., Toro L., F. Vegliò, Metal speciation and pH effect on Pb, Cu, Zn and Cd biosorption onto *Sphaerotilus natans*: Langmuir-type empirical model. Wat Res 37(2003):627–633.

Park D., Yun Y.S., Park J.M. (2010) The past, present, and future trends of biosorption. Biotechnol Bioprocess Eng 15:86–102.

Parvathi K., Nagendran R., Nareshkumar R. (2007) Lead biosorption onto waste beer yeast by-product, a means to decontaminate effluent generated from battery manufacturing industry. E J Biotechnol 10(1). Available from Internet: http://www.ejbiotechnology.info/content/vol10/issue1/full/13/index.html. Accessed 14 June 2015.

Turner B.F., Fein J.B. (2006) Protofit: a program for determining surface protonation constants from titration data. Comput Geociences 32:1344–1356.

Valix M., Tang J.Y., Malik R. (2001) Heavy metal tolerance of fungi. Miner Eng 14(5):499–505.

Worch E. (2012) Adsorption technology in water treatment. Fundamentals, processes and modeling. Walter de Gruyter GmbH & Co. KG, Berlin/Boston.

Xu J., Zhang H., Zhang J., Kim E.J. (2014) Capture of toxic radioactive and heavy metal ions from water by using titanate nanofibers. J Alloys Compd 614:389–393.

Xu P., Zeng G.M., Huang D.L., Lai C., Zhao M.H., Wei Z., Li N.J., Huang C., Xie G.X. (2012) Adsorption of Pb(II) by iron oxide nanoparticles immobilized *Phanerochaete chrysosporium*: Equilibrium, kinetic, thermodynamic and mechanisms analysis. Chem Eng J 203:423–431.

Zaidi A., Oves M., Ahmad E., Khan M.S. (2011) Importance of free-living fungi in heavy metal remediation. In: Biomanagement of metal-contaminated soils, M.S. Khan, A. Zaidi, R.Goel, J. Musarrat (eds.). Environmental pollution 20. Springer. The Netherlands, 479-495 pp.

CHAPTER 6

Effect of selenite on the morpholoy and respiratory activity of *Phanerochaete chrysosporium* biofilms

A modified version of this chapter was published as:

E.J. Espinosa-Ortiz, Y. Pechaud, E. Lauchnor, E. Rene, R. Gerlach, B.M. Peyton, E.D. van Hullebusch, P.N.L. Lens (2016) Effect of selenite on the morphology and respiratory activity of *Phanerochaete chrysosporium biofilms*. Bioresource Technology. Doi:10.1016/j.biortech.2016.02.074.

Abstract

The temporal and spatial effects of selenite (SeO_3^{2-}) on the physical properties and respiratory activity of *Phanerochaete chrysosporium* biofilms, grown in flow-cell reactors, were investigated using oxygen microsensors and confocal laser scanning microscopy (CLSM) imaging. Exposure of the biofilm to a SeO_3^{2-} load of 1.67 mg Se L^{-1} h^{-1} (10 mg Se L^{-1} influent concentration), for 24 h, resulted in a 20% reduction of the O_2 flux, followed by a ~10% decrease in the glucose consumption rate. Long-term exposure (4 days) to SeO_3^{2-} influenced the architecture of the biofilm, by creating a more compact and dense hyphal arrangement resulting in a decrease of biofilm thickness compared to fungal biofilms grown without SeO_3^{2-}. To the best of our knowledge, this is the first time that the effect of SeO_3^{2-} on the aerobic respiratory activity on fungal biofilms is described.

Key words: Fungal biofilm, *Phanerochaete chrysosporium*, selenium, oxygen profiles, microelectrodes

6.1 Introduction

Selenium (Se), an element belonging to group 16 of the periodic table (chalcogens), plays a dual role as an essential micronutrient at low concentrations, but it can also be toxic to organisms at higher concentrations (Lenz and Lens, 2009). An overdose of Se disrupts the integrity of proteins and decreases cellular enzymatic activity, leading to severe consequences for human health, including dermal, respiratory and neurological damage (USHHS, 2003). Se is of significant research interest from an environmental viewpoint due to its toxicity in natural water bodies. In particular, the water-soluble oxyanions selenite (SeO_3^{2-}) and selenate (SeO_4^{2-}), mainly related to agricultural activities, mining, chemical, textile, photographic and electronic industries, has shown to cause widespread impacts on aquatic life, leading to a cascade of detrimental bioaccumulation (Chasteen and Bentley, 2003; Lemly, 2004; Lenz and Lens, 2009).

The use of biological agents to remove pollutants from water can be an efficient, economic and environmentally friendly alternative to conventional physicochemical methods, *e.g.* chemical precipitation, catalytic reduction and ion exchange. Different microorganisms have the ability to convert SeO_3^{2-} and SeO_4^{2-} into insoluble forms, mainly elemental Se (Se^0). The biomineralization of Se has been used for bioremediation of soils and wastewater treatment, with the potential to recover Se biominerals (Nancharaiah and Lens, 2015). Most of the reported biotechnologies for the treatment of Se polluted effluents are based on bacterial metabolism (Nancharaiah and Lens, 2015), although the use of fungi is also promising due to their ability to grow under acidic conditions, ease of handling and its inherent capacity to produce high amounts of enzymes. The latter property is of particular importance since the reduction of metal/metalloid ions and the formation of their elemental nanoparticles is attributed to an enzymatic reduction process (Zhang et al., 2011).

Uptake and volatilization are the main Se coping mechanisms reported for fungi (Gharieb et al., 1995). Some fungal strains, including *Alternaria alteranata* (Sarkar et al., 2011), *Lentinula edodes* (Vetchinkina et al., 2013) and *Phanerochaete chrysosporium* (Espinosa-Ortiz et al., 2015), have demonstrated to biomineralize Se^0 in the nano-size range. These bio-nanomaterials possess improved properties compared to the bulk material, making them a desirable product for use in photocells, semiconductor rectifiers and medical applications (Beheshti et al., 2013). Among the different fungi tested, most of them have been identified as SeO_3^{2-}-reducing organisms, but the reduction of SeO_4^{2-} to Se^0 has only been achieved with the fungal extract of *A. alternata* (Sarkar et al., 2011).

Previous studies on Se-reducing fungal strains were mostly performed using dispersed filamentous or self-immobilized (pellets) cultures (Espinosa-Ortiz et al., 2015; Gharieb et al., 1995). However, the morphological and inhibiting effects of Se oxyanions on fungal biofilms have not yet been reported. Biofilms, as naturally occurring communities of microorganisms, play key roles in the biogeochemical cycling of toxic elements in aquatic systems (van Hullebusch et al., 2003). Biofilms are complex, heterogeneously structured microbial communities attached to a substratum and enclosed in a self-produced extracellular matrix (Flemming and Wingender, 2010). The main differences between attached and dispersed microbial growth arise from the physical structure, leading to different mass transfer patterns, and substrate and nutrient concentration gradients resulting in significant physiological and respiratory heterogeneity.

Biofilms have been suggested to be more productive and metabolically active than planktonic cells, or the dispersed mycelium in the case of fungi, due to high concentrations of active enzymes and extracellular proteins (Ramage et al., 2012). The response of biofilms to inhibitory compounds has been addressed using a number of methods based on indicative targets, *e.g.* respiratory and enzymatic activities, cell growth and cell viability. The use of microsensors to estimate the respiratory activity (O_2 uptake changes) in biofilms exposed to inhibitors has been well established and it has been widely used as a method to assess toxicity and inhibition (Lauchnor and Semprini, 2013; Hou et al., 2014). Therefore, we targeted fungi in a biofilm mode of growth to observe their response to a specific inhibitor (SeO_3^{2-}).

In this paper, we report for the first time the effects of SeO_3^{2-} on *P. chrysosporium*, a Se-reducing organism (Espinosa-Ortiz et al., 2015), when grown as a biofilm in a flow-cell reactor. Experiments were performed first to assess the effects of SeO_3^{2-} exposure on developed biofilms (short-term experiments) and secondly to assess the influence of SeO_3^{2-} on biofilm development (long-term experiments). O_2 microsensor measurements and confocal laser scanning microscopy (CLSM) imaging were employed to visualize the temporal and spatial changes induced by the presence of SeO_3^{2-}. O_2 consumption rates (biofilm activity) and the physical properties of the biofilms, *i.e.* porosity, density, were then estimated based on the O_2 profiles. The changes in the biofilm architecture caused by SeO_3^{2-} were directly observed using CLSM.

6.2 Materials and methods

6.2.1 Fungal strain and culturing conditions

The white-rot fungus *P. chrysosporium* (ATCC 24725) was used throughout the study. The fungal strain was cultured for 3 days at 37 °C on potato dextrose agar slants. Sub-cultures were prepared as required and maintained at 4 °C. A fungal spore suspension, used as the inoculum for subsequent biofilm cultures, was prepared by suspending the culture from one agar slant in 50 mL liquid medium in an Erlenmeyer flask. The liquid medium contained (g L^{-1}): glucose, 2.5; KH_2PO_4, 2; $MgSO_4 \cdot 7H_2O$, 0.5; NH_4Cl, 0.1; $CaCl_2 \cdot 2H_2O$, 0.1; thiamine, 0.001; and 5 mL of a trace element solution as described by Tien and Kirk (1988). The initial pH was adjusted to 4.5 with 1 M HCl, and the medium was sterilized at 123 kPa and at 121 °C for 30 min.

6.2.2 Biofilm growth and exposure experiments

Biofilms of *P. chrysosporium* were cultivated for 4 days in flow-cell reactors (15 cm × 3.8 cm × 3.5 cm) run in parallel in the presence (Se-treated) or absence (untreated) of SeO_3^{2-} (Figure 6.1). Each reactor had a working volume of 0.12 L and a substratum surface area of 57 cm^2. The reactors were filled with medium (2.5 g glucose L^{-1}, pH 4.5, T 30 °C) and the fungus was allowed to grow in batch conditions for one day to facilitate the attachment of the cells. The reactors were then supplied with influent at a constant flow rate (Q_i = 0.02 L h^{-1}), for 3 days, to attain a nominal hydraulic retention time (HRT_n) of 6 h. The effluent was recycled (Q_r = 0.09 L h^{-1}, recirculation to influent ratio = 4.5:1) to provide well-mixed conditions in the reactor. Thus, an actual hydraulic retention time (HRT_a) of 1.1 h was maintained. The substrate loading rate was maintained constant at 0.08 g glucose L^{-1} h^{-1}, while the Se influent concentrations was either 10 or 20 mg Se L^{-1}, corresponding to 1.67 or 3.3 mg Se L^{-1} h^{-1}, respectively. The influent concentrations of glucose and Se reported were measured in the feed medium prior to being mixed in the inlet of the reactors with the recirculation flow (sampling port 1, Figure 6.1); these values were used to estimate the removal rates and efficiencies in the reactor (see Section 2.5). Table 6.1 summarizes the operational parameters used in the reactors during continuous mode of operation. Biofilms were cultivated in closed reactors using atmospheric air as headspace. Whenever required, the reactors were opened for the measurement of O_2 profiles and care was taken to minimize the potential of external contamination. Microscopic imaging revealed no indication of bacterial or archaeal growth in the flow-cell reactors.

Short- and long-term SeO_3^{2-} exposure experiments were performed to assess the spatial and temporal effects on the fungal biofilms, by measuring the O_2 concentration profiles at different times and locations within the biofilms:

(1) *Short-term SeO_3^{2-} exposure tests.* Untreated 4 days-old biofilms were exposed to SeO_3^{2-} (10 mg Se L^{-1}, 1.67 mg Se L^{-1} h^{-1}) for 3 h and 24 h. The O_2 profiles were determined *in situ* before the biofilms were exposed to SeO_3^{2-} (untreated) and again after 3 h and 24 h of SeO_3^{2-} exposure. The results from this experiment were used to determine the reduction in O_2 consumption rate within the biofilm upon SeO_3^{2-} exposure.

(2) *Long-term SeO_3^{2-} exposure tests.* Biofilms were cultivated for 4 days either in the absence or presence of SeO_3^{2-}. The response of the biofilms to (a) low (10 mg Se L^{-1}, 1.67 mg Se L^{-1} h^{-1}), and (b) high (20 mg Se L^{-1}, 3.3 mg Se L^{-1} h^{-1}) influent concentrations was determined. The SeO_3^{2-} concentration was doubled in experiment 2b to delineate the influence of a different SeO_3^{2-} load on the biofilm. A physical characterization of these biofilms was performed with CLSM. O_2 profiles were measured within the biofilms at the end of the incubation period (4 days) of each experiment.

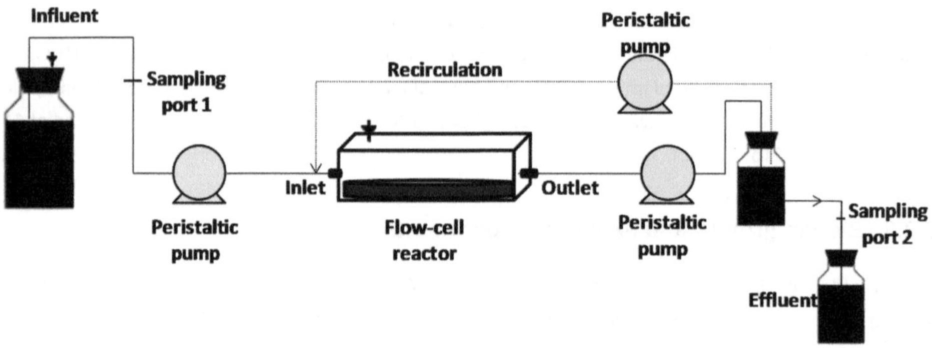

Figure 6.1 Schematic of the flow -cell system.

Table 6.1 Operational parameters of the flow-cell reactors.

Parameter	Value
Working volume	0.120 L
Substratum surface area	57 cm^2
Nominal hydraulic retention time*	6 h
Actual hydraulic retention time**	1.1 h
Influent flow rate	0.02 L h^{-1}
Recirculation flow rate	0.09 L h^{-1}
Recirculation ratio***	4.5
pH of feed	4.5
Temperature	30 °C
Substrate loading rate	0.08 g glucose L^{-1} h^{-1}
Se loading rate†	1.67 or 3.3 mg Se L^{-1} h^{-1}
Se influent concentration†	10 or 20 mg Se L^{-1}

Note:

*$HRT_n = \frac{Working\ volume\ (L)}{Influent\ flow\ rate\ (mg\ L^{-1})}$

**$HRT_a = \frac{Working\ volume\ (L)}{Influent\ flow\ rate\ (mg\ L^{-1}) + Recirculation\ flow\ rate\ (mg\ L^{-1})}$

***$S = \frac{Recirculation\ flow\ rate\ (mg\ L^{-1})}{Influent\ flow\ rate\ (mg\ L^{-1})}$

† Corresponds to the Se concentration or Se loading rate of the medium prior to being mixed in the inlet with the recirculation flow.

6.2.3 Microsensor measurements

Microsensors for O_2 measurements were used to determine the influence of SeO_3^{2-} on O_2 concentrations within the *P. chrysosporium* biofilms. O_2 profiles over the depth of the biofilm at three different locations in the bioreactor were measured using a Clark-type O_2 microsensor with a tip diameter of 8-12 μm (Unisense, Denmark), having a detection limit of 0.01 mg O_2 L^{-1}. Measurements were made *in situ*, by maintaining the flow continuously in the bioreactors, at room temperature and under atmospheric O_2 conditions. Positioning of the microsensors was done using a micromanipulator fitted with a motor controller, and the sensor output data was collected online using the SensorTrace Pro software (Unisense, Denmark). O_2 profiles were acquired in triplicate at each location. The profiles obtained were normalized by dividing the O_2 concentration measured at depth *Z* by the O_2 concentration in the bulk solution. The observation of similar O_2 profiles for each triplicate suggested that pseudo-steady-state O_2 uptake was reached. Based on the O_2 profiles, the biofilm surface was determined according to the procedure outlined by Wäsche et al. (2002).

6.2.3.1 Estimation of diffusion coefficients

The effective diffusion coefficients of O_2 in the biofilms were estimated from the O_2 profiles. Obeying Fick's law, and considering one-dimensional diffusion gradients in the biofilm, the O_2 flux in the boundary layer between the bulk solution and the biofilm was estimated according to Equation 1. The O_2 flux in the biofilm was determined using Equation 2 (Lewandowski et al., 1991):

$$Flux, J_w\ (mg\ cm^{-2} s^{-1}) = -D_w \left(\frac{dC}{dz}\right)_{External} \quad \text{Eq. (1)}$$

$$Flux, J_f\ (mg\ cm^{-2} s^{-1}) = -D_f \left(\frac{dC}{dz}\right)_{Internal} \quad \text{Eq. (2)}$$

where D_w (cm^2 s^{-1}) is the O_2 diffusion coefficient in water and D_f (cm^2 s^{-1}) is the diffusion coefficient within the biofilm. *dC/dz* is the gradient O_2 concentration measured with the microsensors at the boundary layer (*external*) or within the biofilm (*internal*).

Assuming the fluxes J_w and J_f are equal at the surface, the diffusion coefficient D_f, can then be calculated according to Equation 3 (Wanner et al., 2006):

$$D_f = D_w \left(\frac{dC}{dz}\right)_{External} / \left(\frac{dC}{dz}\right)_{Internal} \qquad \text{Eq. (3)}$$

The relative diffusivity (f_D), which is defined as the relation between D_f and D_w, was estimated using Equation 4:

$$D_f = f_D D_w \qquad \text{Eq. (4)}$$

6.2.3.2 Estimation of biofilm physical properties based on diffusion coefficients and O_2 inhibition

Zhang and Bishop (1994) developed a random porous cluster model, assuming that the microorganisms are randomly located in a fraction of the available sites in the biofilm, avoiding multiple occupancies. Based on this model, the porosity of the biofilm (θ) can be calculated according to Equation 5:

$$\theta = \sqrt[3]{f_D} \qquad \text{Eq. (5)}$$

The density (ρ, kg m^{-3}) of the biofilm can also be approximately calculated based on its correlation to the relative diffusivity, f_D (Fan et al., 1990), according to Equation 6. It should be noted that this correlation is based on the experimental data obtained for different types of biofilms (including fungi), substrates and temperatures.

$$\frac{D_f}{D_w} = 1 - \frac{0.43\ \rho^{0.92}}{11.19+0.27\ \rho^{0.99}} \qquad \text{Eq. (6)}$$

The global inhibition, I_j (%), *i.e.* the reduction in O_2 flux into the biofilms, and thus O_2 consumption rate, in the presence of SeO_3^{2-} relative to the control biofilms, was estimated according to Equation 7 (Zhou et al., 2011).

$$I_j = 1 - \left(\frac{J_{W\ Se-Treated}}{J_{W\ Untreated}}\right) \qquad \text{Eq. (7)}$$

6.2.4 Biofilm sectioning and imaging

Samples from biofilms grown for 4 days in the absence or presence of SeO_3^{2-} (20 mg Se L^{-1}, 3.3 mg Se L^{-1} h^{-1}) were chosen randomly and stained with the fluorescent probe FUN 1 (Molecular Probes), as previously described elsewhere (Villa et al., 2011). Briefly stating, the biofilms were stained by covering the biofilm samples with 30 µM FUN 1 in phosphate buffered saline (PBS) solution at 30 °C for 30 min in the dark. The biofilms were then rinsed with PBS solution. The stained samples were covered with optimum cutting temperature formulation (OCT, Tissue-Tek, USA) and frozen with dry ice. The frozen core was embedded in more OCT, and then sectioned at -20 °C using a Leica CM1850 cryostat. Cryo-sections (7 µm thick) were visualized with a Nikon Eclipse E800 microscope in transmission mode using differential interference contrast optics. Images were processed using the software MetaMorph (Molecular Devices, Downington, PA). Biofilm architecture of the stained biofilms was visualized using CLSM (Leica TCS-SP2 AOBS). The software Imaris (Bitplane Scientific Software, Zurich, Switzerland) was used for processing the images.

6.2.5 Analytical methods

Influent and effluent samples were collected twice per day and centrifuged at 4700 g for 15 min, and the supernatant was used for analysis. Glucose concentrations were determined using the dinitrosalicylic acid method (Miller, 1959) with D-glucose as the standard. The total Se concentration was measured using ICP-MS (Agilent 7500ce) after preserving the samples with 5% HNO_3 (Espinosa-Ortiz et al., 2015).

6.2.5.1 Performance parameters of the flow-cell reactor

The performance of the flow-cell reactor was determined by calculating the glucose and Se removal rates according to Equation 8, while the removal efficiency (E) was estimated using Equation 9:

$$Removal\ rate, R_i(mg\ L^{-1}\ h^{-1}) = \frac{Q\ [C_0 - C_t]}{V} \quad \text{Eq. (8)}$$

$$Removal\ efficiency, E_i(\%) = \frac{C_0 - C_t}{C_0} \quad \text{Eq. (9)}$$

where Q (L h^{-1}) is the influent flow rate, C_0 (mg L^{-1}) is the influent concentration, C_t (mg L^{-1}) is the effluent concentration measured at time t, and V (L) is the working volume of the reactor (0.120 L). The subscript i in the equations indicates either Se or glucose.

6.2.5.2 Statistical analysis

The substrate and Se removal rates by *P. chrysosporium* biofilms are reported in this study. The mean and standard deviation values were calculated for the long term experiments when steady state conditions were attained over the 4 days of incubation. The effect caused by the presence of SeO_3^{2-} on the E_i was evaluated by performing analysis of variance (ANOVA) at $P_{value} \leq 0.05$.

6.3 Results

6.3.1 Influence of short-term SeO_3^{2-} exposure on *P. chrysosporium* biofilm activity

Short-term experiments were performed to ascertain the temporal effects of SeO_3^{2-} on the *P. chrysosporium* biofilms, which allows to estimate the inhibition of the global O_2 consumption. Biofilms were grown for 4 days in the absence of SeO_3^{2-} and then exposed to SeO_3^{2-} (10 mg Se L^{-1}, 1.67 mg Se L^{-1} h^{-1}), for 3 and 24 h (Figure 6.2). O_2 concentration profiles were determined at different locations in the biofilm, before and after exposure to SeO_3^{2-} (Figure 6.3). The O_2 concentration in the bulk solution varied between 7.0 and 7.5 mg O_2 L^{-1}. Triplicates of the O_2 profiles were acquired for each location; similar profiles were obtained during the measurements of each location (data not shown), which suggested a pseudo-steady-state O_2 uptake by the biofilm.

Table 6.2 summarizes the parameters obtained from the O_2 profiles. Before and after SeO_3^{2-} exposure, O_2 concentrations reached levels below the detection limit in the *P. chrysosporium* biofilm indicating the presence of anoxic zones within the biofilm (Figure 6.2). Differences in the O_2 profiles observed over time revealed the influence of SeO_3^{2-} on the O_2 flux into the biofilm. While the differences between *lj* values observed were not statistically different (P_{value} = 0.13), a decrease in O_2 consumption rate of 11.5 (± 6.5)% after SeO_3^{2-} exposure for 3 h and 20 (± 5.3)% upon exposure for 24 h was observed. The rates of glucose consumption ($R_{glucose}$) were estimated to be 0.052, 0.049 and 0.047 g glucose L^{-1} h^{-1} after 0, 3 and 24 h of SeO_3^{2-} treatment, respectively, corresponding to a 6% and 10% decrease in $R_{glucose}$. The rates of Se removal (R_{Se}) were calculated to be 0.17 mg Se L^{-1} h^{-1} after 3 h and 0.22 mg Se L^{-1} h^{-1} after 24 h, corresponding to E_{Se} values of 12% and 15.5%, respectively.

As shown in Figure 6.2, an orange-red coloration developed in the biofilm over time during SeO_3^{2-} exposure. After 3 h, a visible coloration was observed, which developed into overt differences in coloration intensity after 24 h. The change of biofilm coloration from white-ivory to orange-red is an indicator of the occurrence of SeO_3^{2-} reduction to red amorphous Se^0 (Espinosa-Ortiz et al., 2015; Sarkar et al., 2011).

Table 6.1 Parameters calculated for *P. chrysosporium* biofilms cultivated in the flow-cell reactors for 4 days in the absence of SeO_3^{2-} (untreated) and then treated with 1.67 mg Se L^{-1} h^{-1} (10 mg Se L^{-1}) as SeO_3^{2-} for 3 and 24 h*.

Incubation	Effective O_2 diffusion coefficient in the biofilm, D_f (cm^2 s^{-1})	Relative diffusivity, f_D**	Porosity, θ**	Density, ρ (kg m^{-3})*	O_2 flux, J_w (mg cm^{-2} s^{-1})**
Untreated	$8.8\times10^{-6}\pm9.9\times10^{-7}$	0.44±0.05	0.76±0.03	35.4±7.2	$4.9\times10^{-7}\pm6.5\times10^{-8}$
Treated for 3 h	$9.9\times10^{-6}\pm1.4\times10^{-6}$	0.49±0.07	0.79±0.04	29.5±7.8	$4.4\times10^{-7}\pm9.0\times10^{-8}$
Treated for 24 h	$7.5\times10^{-6}\pm7.3\times10^{-7}$	0.37±0.04	0.72±0.02	46.5±7.2	$3.9\times10^{-7}\pm7.8\times10^{-8}$

Note:
* Triplicates of the measured profiles at three different locations were used for calculations (n=9).
**Parameters were calculated based on the estimated effective diffusion coefficients from the experimental data according to Equations 4, 5, 6 as well as 1 and 2.

The global biofilm activity was also measured in terms of substrate consumption rate. The untreated biofilm showed a $R_{glucose}$ value of 0.054 (± 0.005) g glucose L^{-1} h^{-1}, corresponding to $E_{glucose}$ of 52 (± 4.8)%, after 4 days of incubation. The Se-treated biofilm showed a $R_{glucose}$ value of 0.044 (± 0.004) g glucose L^{-1} h^{-1} ($E_{glucose}$ 41 ± 4.3%). The difference in $E_{glucose}$ between the untreated and Se-treated biofilms was statistically significant (P_{value} = 0.041), demonstrating that the incubation with SeO_3^{2-} decreases $E_{glucose}$ in the biofilms. The removal of Se by the biofilm was estimated for the Se-treated biofilm, with an average R_{Se} of 0.32 (± 0.02) mg Se L^{-1} h^{-1} (E_{Se} 20 ± 0.4%), after 4 days of incubation. At the end of the incubation period (4 days), the pH of the effluent decreased from 4.5 to 3.8 (± 0.02) for both Se-treated and untreated biofilms.

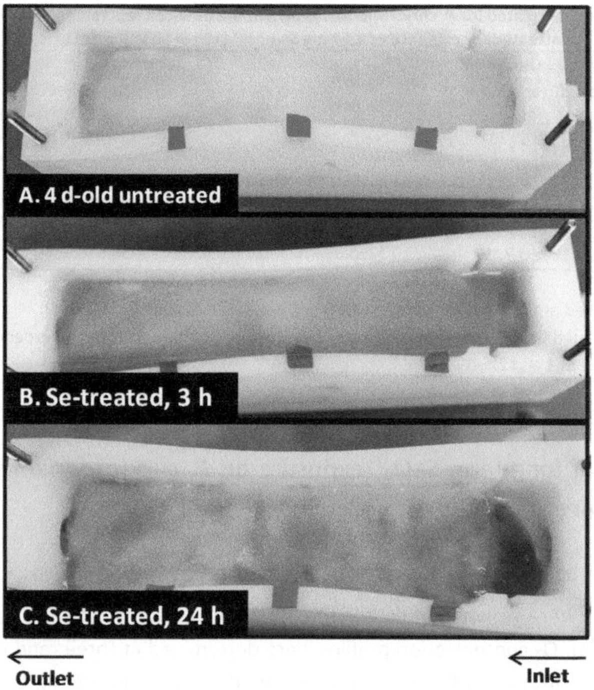

Figure 6.2 ***P. chrysosporium*** **biofilms cultivated for 4 days in the absence of SeO_3^{2-} (A), and then treated with 1.67 mg Se L^{-1} h^{-1} (10 mg Se L^{-1}) as SeO_3^{2-} for 3h (B) and 24 h (C).**

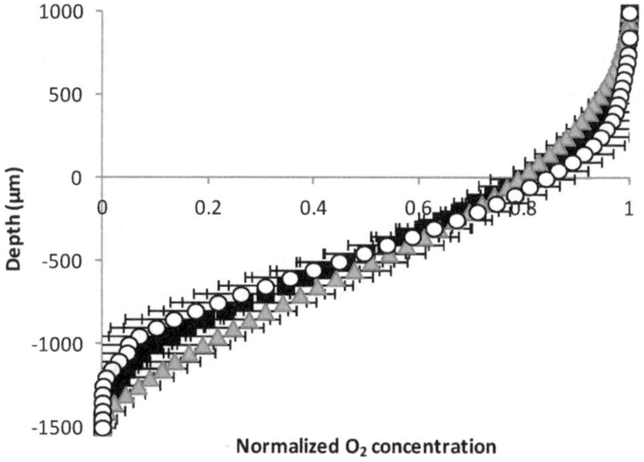

Figure 6.3 O_2 profiles measured at three different locations in ***P. chrysosporium*** **biofilms cultivated for 4 days in the absence of SeO_3^{2-} (■) and then treated with 1.67 mg Se L^{-1} h^{-1} (10 mg Se L^{-1}) as SeO_3^{2-} for 3(▲) and 24 h (○).**

Table 6.3 Parameters calculated for *P. chrysosporium* grown in the flow-cell reactors for 4 days in the absence (untreated) or presence of 1.67 mg Se L^{-1} h^{-1} (10 mg Se L^{-1}) as SeO_3^{2-*}.

Incubation	Effective O_2 diffusion coefficient in the biofilm, D_f (cm^2 s^{-1})	Relative diffusivity, f_D**	Porosity, θ**	Density, ρ (kg m^{-3})**	O_2 flux, J_w (mg cm^{-2} s^{-1})**
Untreated	$8.1\times10^{-6}\pm4.8\times10^{-7}$	0.41±0.02	0.74±0.02	39.7±4.5	$5.2\times10^{-7}\pm3.0\times10^{-8}$
Treated	$6.6\times10^{-6}\pm3.7\times10^{-7}$	0.32±0.03	0.69±0.03	56.6±6.9	$4.2\times10^{-7}\pm2.3\times10^{-8}$

Note:

* Triplicates of the measured profiles at three different locations were used for calculations (n=9).

** Parameters were calculated based on the estimated effective diffusion coefficients from the experimental data according to Equations 4, 5, 6 as well as 1 and 2.

6.3.2 Influence of long-term SeO_3^{2-} exposure on *P. chrysosporium* biofilm activity and structure

6.3.2.1 Response to low Se influent concentration (10 mg Se L^{-1})

Fungal biofilms were grown for 4 days in the presence or absence of SeO_3^{2-} (10 mg L^{-1}, 1.67 mg Se L^{-1} h^{-1}) (Figure 6.4A). O_2 concentration profiles were determined at three random locations in the biofilms. The O_2 concentration in the bulk solution varied from 7.0 to 7.5 mg O_2 L^{-1} (at 23 °C). The profiles were normalized as described in Section 6.2.3. Average measurements for the untreated and Se-treated biofilms are shown in Figure 6.5A. Irrespective of the growth conditions, O_2 concentrations decreased to below the microsensor detection limit indicating that the biofilms developed anoxic zones. Table 6.3 provides the parameters obtained from the O_2 profile measurements in the Se-treated and untreated biofilms. These results show the effective O_2 diffusion coefficient is ~20% lower for the Se-treated biofilms. Based on Equation 6, the biofilm density of the Se-treated biofilms was ~30% higher than the density of the untreated biofilms. Based on Equation 5, the porosity of the untreated biofilm decreased by ~7% compared to the untreated biofilms.

Upon 4 days of SeO_3^{2-} exposure, the biofilm showed an orange-red coloration a, suggesting the biomineralization of Se^0. A transversal cut (cross-section) of the biofilm (Figure 6.4B) generally showed an intensification of this orange-red coloration from the top to the bottom of the biofilm. The upper zone of the biofilm showed either the white-ivory characteristic color of *P. chrysosporium* biofilms or a slight red coloration. Image analysis of the cross-section of the biofilm treated with SeO_3^{2-} (10 mg Se L^{-1}, 1.67 mg Se L^{-1} h^{-1}) (Figure 6.4B) indicated a maximal thickness of 3500 μm, providing an indication of the actual thickness of the biofilms grown in this study.

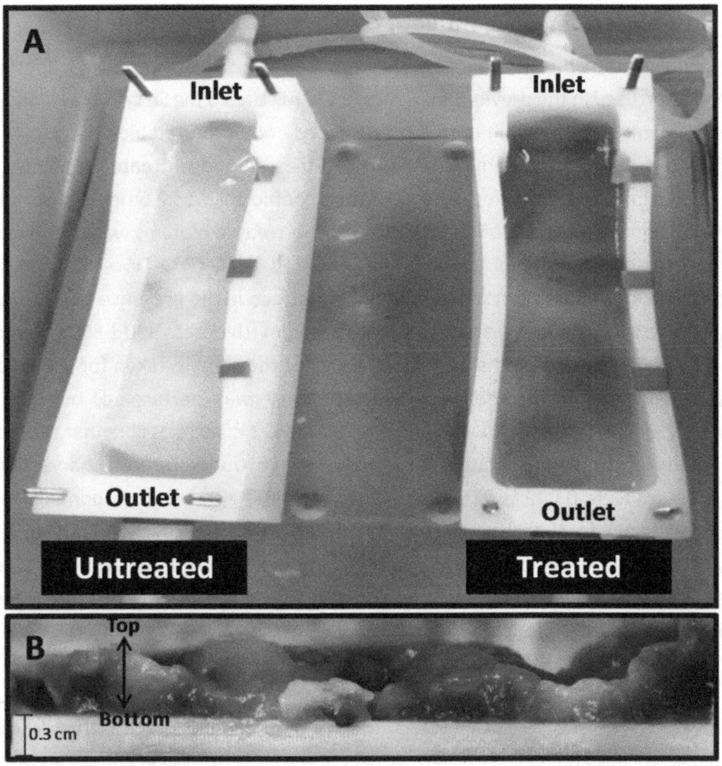

Figure 6.4. (A) Biofilms of *P. chrysosporium* grown in the flow-cell reactors for 4 days in the absence or presence of 1.67 mg Se L^{-1} h^{-1} (10 mg Se L^{-1}) as SeO_3^{2-}. (B) Cross-sectional view of a 4 day-old biofilm grown in the presence of 1.67 mg Se L^{-1} h^{-1} (10 mg Se L^{-1}) as SeO_3^{2-}.

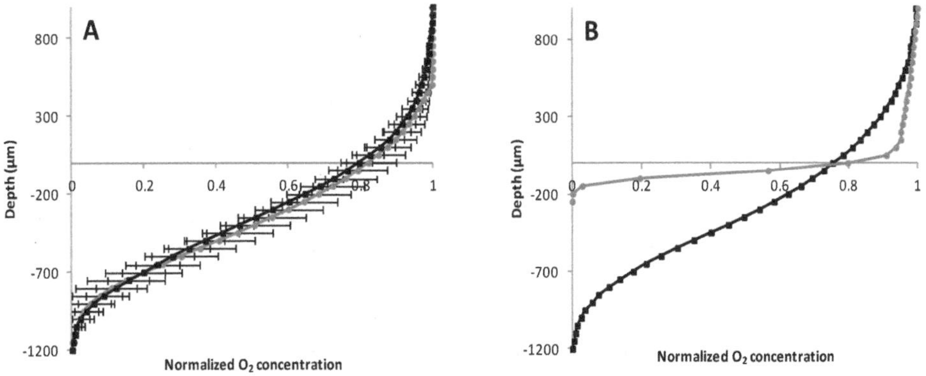

Figure 6.5. Average oxygen profiles measured at: (A) different locations in biofilms grown in flow-cell reactors for 4 days in the absence (■) or presence of 1.67 mg Se L^{-1} h^{-1} (10 mg Se L^{-1}) as SeO_3^{2-} (●); (B) O2 profiles measured in one location in biofilms grown in the flow-cell reactors for 4 days in the absence (■) or presence of 3.3 mg Se L^{-1} h^{-1} (20 mg Se L^{-1}) as SeO_3^{2-} (●).

6.3.2.2 Response to high Se influent concentration (20 mg Se L^{-1})

O_2 concentration profiles (Figure 6.5B) were measured after 4 days of incubation with twice the Se concentration used in the previous experiment (20 mg L^{-1}, 3.3 mg Se L^{-1} h^{-1}). O_2 concentrations below the detection limit were noted in both the Se-treated and untreated biofilms (Figure 6.5B). Table 6.4 summarizes the calculated parameters for the Se-treated and untreated biofilms based on the O_2 profiles. The estimate for the effective O_2 diffusion coefficient was 2.6 times lower for the Se-treated biofilm. Using Equation 7, the biofilm density of the Se-treated biofilms was calculated to be 3.3 times higher than for the density of biofilms not exposed to SeO_3^{2-}. Cryo-sectioning combined with microscopy indicated that *P. chrysosporium* biofilm formed in the presence of SeO_3^{2-} was considerably thinner (thickness 79 ± 40 μm) than the untreated biofilm (thickness 390 ± 85 μm) equivalent to a 80% reduction in biofilm thickness. It should be noted that the sample taken for the cryo-sectioning was randomly selected from the area where O_2 profiling was performed, but due to the biofilm heterogeneity, actual thickness could have varied. Anew, it is also possible that the biofilm might have contracted during sample preparation for microscopy (*i.e.* during freezing, harvesting, staining, etc.). The $R_{glucose}$ values were 0.049 (± 0.004) and 0.034 (± 0.003) g L^{-1} h^{-1}, corresponding to $E_{glucose}$ values of 46 (± 4) and 32 (± 4)% for the untreated and Se-treated biofilms, respectively. The differences between the $R_{glucose}$ (P_{value} = 0.006) and $E_{glucose}$ (P_{value} = 0.012) between the untreated and Se-treated biofilms were statistically significant, confirming the inhibition of substrate consumption in the biofilms due to SeO_3^{2-} exposure. The R_{Se} was 0.30 (± 0.17) mg Se L^{-1} h^{-1} (E_{Se} = 6.6 ± 2.5%).

Table 6.4 Parameters calculated for biofilms of *P. chrysosporium* grown in the flow-cell reactors for 4 days in the absence (untreated) or presence of 3.3 mg Se L^{-1} h^{-1} (20 mg Se L^{-1}) as SeO_3^{2-} *.

Incubation	Effective O_2 diffusion coefficient in the biofilm, D_f (cm^2 s^{-1})	Relative diffusivity, f_D**	Porosity, θ**	Density, ρ (kg m^3)**	O_2 flux, J_w (mg cm^{-2} s^{-1})**
Untreated	8.9×10^{-6}±1.0×10^{-7}	0.44±0.02	0.76±0.02	35.3±2.6	5.2×10^{-7}±5.8×10^{-8}
Treated	3.4×10^{-6}±2.0×10^{-7}	0.17±0.03	0.55±0.03	117.6±21.5	9.5×10^{-7}±5.4×10^{-8}

Note:

* Triplicates of the measured profiles at three different locations were used for calculations (n=9).

** Parameters were calculated based on the estimated effective diffusion coefficients from the experimental data according to Equations 4, 5, 6 as well as 1 and 2.

The architecture of both untreated and Se-treated biofilms (SeO_3^{2-} exposure for 4 days, 20 mg L^{-1}, 3.3 mg Se L^{-1} h^{-1}) was visualized using CLSM. The images of the untreated and Se-treated biofilms (Figure 6.6) suggest that the structure of the biofilm was affected by the presence of SeO_3^{2-}. While both, untreated and Se-treated biofilms appeared to be composed of an elaborate network of hyphal elements, a more compacted and dense structure was observed for the Se-treated samples. This confirms the densification of the Se-treated biofilms as suggested by the estimated overall biofilm density based on Equation 6 and the decrease of the effective O_2 diffusion coefficient in the Se-treated biofilms (Table 6.4). Moreover, side views of the three-dimensional reconstructed CLSM images indicated denser and ~50% shorter hyphae in the Se-exposed biofilms (Figures 6.6B and D).

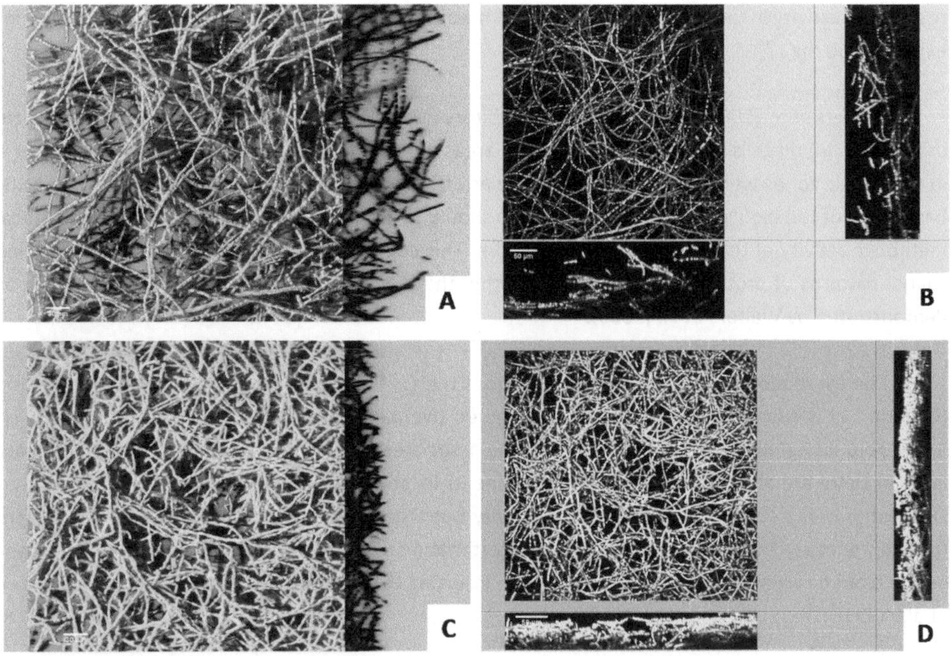

Figure 6.6 Confocal scanning laser microscopy (CSLM) images of *P. chrysosporium* biofilms grown in the flow-cell reactors for 4 days in the (A,B) absence and (C,D) presence of 3.3 mg Se L^{-1} h^{-1} (20 mg Se L^{-1}) as SeO_3^{2-}. Bars represent 30 µm for A and C, and 50 µm for B and D.

6.4 Discussion

6.4.1 Inhibition of developed *P. chrysosporium* biofilms by SeO_3^{2-}

To the best of our knowledge, this is the first time that the effect of SeO_3^{2-} on the respiratory activity and morphology of fungal biofilms is reported. Based on the short-term experiments, the exposure of SeO_3^{2-} (10 mg L^{-1}, 1.67 mg L^{-1} h^{-1}) was inhibitory to *P. chrysosporium.* According to the I_j values obtained, up to 11.5% inhibition of the O_2 respiration occurred within 3 h of exposure time. A diminution in the fungal activity was further confirmed by a 6% decrease in the $R_{glucose}$ of the biofilm. After 24 h of SeO_3^{2-} exposure, a more pronounced inhibition of the fungal activity was observed. The Ij was ~25 %, while the $R_{glucose}$ decreased by ~10% compared to the untreated biofilm. Besides, varying inhibition levels were observed in different sections of the biofilm which was attributed to the heterogeneous nature of the biofilm.

The inhibition of the O_2 respiration and the decrease in the $R_{glucose}$ of the biofilm indicate that SeO_3^{2-} exposure is inhibitory to the activity of *P. chrysosporium*. Even though O_2 profiles varied across the length of the reactor, which was attributed to the heterogeneous nature of the biofilm, a clear inhibition in O_2 consumption was observed in the presence of SeO_3^{2-}. I_j did not vary significantly over time, suggesting that the inhibitory response of *P. chrysosporium* occurs within 3 h of SeO_3^{2-} exposure. This agrees with observations by Zhou et al. (2011) who observed a 29% inhibition of O_2 consumption within 1 h of exposure of wastewater biofilms to Zn^{2+} (5 mg L^{-1}). Analogous responses were observed in aerobic wastewater biofilms exposed to ZnO nanoparticles (Hou et al., 2014), wherein the O_2

respiration activity in the outer layer of the biofilm was inhibited (27%) within 2 h of ZnO nanoparticles exposure (50 mg L^{-1}).

Walters et al. (2003) showed stronger inhibitory effects of antibiotics (10 μg mL^{-1} of ciprofloxacin or tobramycin) on cells in the more oxygenated areas of *P. aeruginosa* biofilms, whereas the bacteria in the anoxic zones were more resistant. This was attributed to low metabolic activity in the anoxic basal layer of the biofilm. Further investigations should be performed in order to determine the local metabolic activity of the *P. chrysosporium* biofilm. This could for instance be achieved by visualizing spatial patterns of protein synthetic activity within the biofilm using epifluorescence microscopy as demonstrated by Walters et al. (2003).

The estimation of the inhibitory effect from the O_2 flux based on Fick's first law of diffusion (Equation 7) is commonly considered to be a good overall estimate of biofilm respiratory activity, although uncertainties in these estimates can be associated with: (i) the estimation of the boundary layer as there are different methods to determine its location (Wäsche et al., 2002), (ii) the fact that O_2 flux not only depends on the microbial activity but also on the physical properties of the biofilm (possibly porosity and density), and (iii) the overestimation of the interfacial flux at the boundary layer as the biofilm might become compressed when inserting the microsensor into the biofilm (Lorenzen et al., 1995).

The inhibitory effect of SeO_3^{2-} on fungal growth has been previously described for unicellular, polymorphic and filamentous fungi (Gharieb et al., 1995; Espinosa-Ortiz et al., 2015). The response of *P. chrysosporium* biofilms to SeO_3^{2-} exposure was similar to the responses observed on *P. chrysosporium* pellets (Espinosa-Ortiz et al., 2015). When grown as pellets in the presence of SeO_3^{2-} (10 mg Se L^{-1}, 10 g glucose L^{-1}, pH 4.5, 30 °C, 12 days incubation), the biomass production of the fungus decreased by ~80% compared to Se-free incubations, while the substrate consumption was limited to ~30%. The change of biofilm coloration from white-ivory to orange-red is an indicator of the occurrence of SeO_3^{2-} reduction to red amorphous Se^0 (Espinosa-Ortiz et al., 2015; Sarkar et al., 2011). The intracellular synthesis of Se^0 from SeO_3^{2-} was demonstrated in the fungal pellets (Espinosa-Ortiz et al., 2015), with formation of spherical particles of amorphous red Se^0 in the nano-size range (30-400 nm). Although chemical characterization of the Se particles in the biofilm was not performed in the present study, the change of coloration upon SeO_3^{2-} exposure to red-orange in the *P. chrysosporium* biofilms strongly suggests that Se^0 formation occurred in the biofilms.

6.4.2 Influence of SeO_3^{2-} on the physical and morphological properties of *P. chrysosporium* biofilms

During long-term exposure experiments, O_2 microsensor measurements (Figure 6.5A) revealed less influence of SeO_3^{2-} on the O_2 consumption activity through the depth of the biofilm after 4 days of exposure (10 mg Se L^{-1}, 1.67 mg Se L^{-1} h^{-1}). O_2 penetrated deeper into SeO_3^{2-}-treated biofilms, suggesting a decrease of the O_2 respiratory activity at the upper layers of the biofilms. This decrease was observed despite the estimated higher biofilm densities, which would otherwise result in a smaller O_2 penetration depth due to higher O_2 consumption per unit volume in the biofilm in the absence of a lower respiratory activity. The increase of biofilm density was calculated to be ~30% compared to the untreated biofilms (Table 6.3), with a ~7% decrease of the biofilm porosity. It is noteworthy to mention that the density and porosity of the biofilms were not measured directly in this study but rather estimated based on the diffusion coefficient estimates derived from the O_2 profiles (Fan et al., 1990; Zhang and Bishop, 1994). For both the 10 and 20 mg Se L^{-1} exposure, smaller diffusion coefficients were estimated compared to the untreated biofilms (Tables 6.3 and 6.4), suggesting high density structures (Hibiya et al., 2004) with low porosities (Zhang and Bishop, 1994). The diffusive transport within biofilms generally decreases with increasing density and decreasing porosity of the biofilm (*e.g.* Fan et al., 1990).

CLSM imaging indicated a more compact biofilm microstructure (Figure 6.6) and a decrease in biofilm thickness in the presence of SeO_3^{2-}, suggesting a densification process and supporting the results obtained through the O_2 profiles. Densification is a recognized mechanism of resistance in biofilms (Mah and O'Toole, 2001). Changes in biofilm structure, including the formation of more compact or dense biofilms, can be a biofilm stress response. The presence of metals induces the formation of stiffer or more compact bacterial biofilms. In a recent study, compaction was induced in *Xilena fastidiosa* when the biofilms were exposed to sub-inhibitory concentrations of Zn^{2+} (Navarrete and De la Fuente, 2014). In yeast biofilms of *Candida albicans* and *C. tropicalis*, sub-inhibitory concentrations of metal ions (Zn^{2+}, Co^{2+}, Cu^{2+}, Ag^{+}, Cd^{2+}, Hg^{2+}) and metalloid oxyanions (AsO_2^- and SeO_3^{2-}) inhibited hyphal formation, leading to overt changes in biofilm structure (Harrison et al., 2007). The effect of SeO_3^{2-} on the fungal morphology has already been demonstrated for *P. chrysosporium* pellets by Espinosa-Ortiz et al. (2015), wherein the authors demonstrated the occurrence of more compact, smaller and denser structures in the presence of Se. Stress due to toxicant exposure can increase the synthesis rate of extracellular polymeric substances (EPS) and/or change the nature and properties of EPS produced, which could lead to spatial restructuring of the biofilms. Dhanjal and Cameotra (2011) observed that SeO_3^{2-} exposure induces the production of large quantities of EPS in *Bacillus* sp., with different composition compared to cells cultivated in the absence of Se. Thus, the presence of SeO_3^{2-} could induce changes in the EPS production of *P. chrysosporium* biofilms and in turn modify the physical properties of the biofilms. Further research to assess the impact on EPS production and composition in fungal biofilms due to SeO_3^{2-} exposure is recommended as one of the next steps.

When biofilm densification occurs the microbial growth inside the biofilm appears to result in an increase of cell density rather than an increase in biofilm thickness (Ramsay et al., 1989). This could explain the noticeable decrease in biofilm thickness (~80%) when treated with 3.3 mg Se L^{-1} h^{-1} (20 mg Se L^{-1}) SeO_3^{2-}. It should be noted that the cryo-sectioning method used in this study to estimate the thickness of the fungal biofilm (Villa et al., 2011) is most commonly used for thin biofilms (<500 µm) and might result in less accurate estimates for thicker biofilms (Yu et al., 1994 and discussed above). Image analysis of the transversal cut of the biofilm treated with 1.67 mg Se L^{-1} h^{-1} (10 mg Se L^{-1}) SeO_3^{2-}, revealed a maximum thickness of 3500 µm, providing an indication of the actual thickness of the biofilms grown in this study. A decrease in biofilm thickness has been reported in biofilms as a response to the presence of toxicants or inhibitors. The thickness of *Candida albicans* biofilms decreased by ~77% when incubated for 72 h in the presence of zosteric acid sodium salt, a well-known anti-biofilm compound (Villa et al., 2011).

The heterogeneity and activity of the biofilm was clearly visible as a result of different intensities of the orange-red coloration, with particularly high color intensity in the region near the reactor's inlet, indicating gradients of Se^0 accumulation and decreasing amounts of Se^0 with distance from the reactor inlet. Moreover, observations made on the cross-section of the 4 day-old Se-treated biofilm (10 mg Se L^{-1}) also indicated gradients of coloration with biofilm depth, with the most intense color at the bottom of the biofilm. This suggests that significant quantities of red amorphous Se^0 accumulated in the deepest layers of the biofilm. Further studies to determine the distribution and speciation of Se species inside the biofilms are suggested, using for example grazing-angle X-ray spectroscopic techniques, X-ray standing wave (XSW) and X-ray absorption near edge structure (XANES) spectroscopy (Templeton et al., 2003), or X-ray fluorescence imaging (XFI) and scanning transmission X-ray microscopy (STXM) (Yang et al., 2016).

The biofilm architecture and structure can also be altered by different factors, including pH, temperature, availability of O_2 and other agents that might cause stress to the microorganisms (Pettit et al., 2010; Kucharíková et al., 2011). This study showed that SeO_3^{2-} also alters the fungal biofilm morphology. The effect of SeO_3^{2-} on the fungal morphology has already been demonstrated for *P. chrysosporium* pellets by Espinosa-Ortiz et al. (2015), who demonstrated the occurrence of more compact, smaller and denser structures in the presence of Se.

6.5 Conclusions

The effects of SeO_3^{2-} on the O_2 respiratory activity and the physical properties of *P. chrysosporium* biofilms were evaluated using O_2 microsensors combined with CLSM imaging. Short-term exposure to SeO_3^{2-} had an inhibitory effect on the respiratory activity of biofilms. The presence of SeO_3^{2-} showed a marked influence on the morphology of the biofilms, leading to the formation of a more compact and dense structures with, on average, shorter fungal hyphae, which is attributed to a mechanism of resistance or a secondary response due to Se^0 biomineralization.

6.6 References

Beheshti, N., Soflaei S., Shakibaie, M., Yazdi, M.H., Ghaffarifar, F., Dalimi, A., Shahverdi, A.R. (2013) Efficacy of biogenic selenium nanoparticles against *Leishmania major*: in vitro and in vivo studies. J Trace Elem Med Biol 27(3):203–207.

Chasteen, T.G., Bentley, R. (2003) Biomethylation of selenium and tellurium: microorganisms and plants. Chem Rev 103:1–25.

Dhanjal, S., Cameotra, S.S. (2011) Selenite stress elicits physiological adaptations in *Bacillus* sp. (strain JS-2). J Microbiol Biotechnol 21:1184–1192.

Espinosa-Ortiz, E.J., Gonzalez-Gil, G., Saikaly, P.E., van Hullebusch, E.D., Lens, P.N.L. (2015) Effects of selenium oxyanions on the white-rot fungus *Phanerochaete chrysosporium*. Appl Microbiol Biotechnol 99:2405–2418.

Fan, L.S., Leyva-Ramos, R., Wisecarver, K.D., Zehner, B.J. (1990) Diffusion of phenol through a biofilm grown on activated carbon particles in a draft-tube three-phase fluidized-bed bioreactor. Biotechnol Bioeng 35:279–286.

Flemming, H.C., Wingender, J. (2010). The biofilm matrix. Nat Rev Microbiol 8:623–633.

Gharieb, M.M., Wilkinson, S.C., Gadd, G.M. (1995) Reduction of selenium oxyanions by unicellular, polymorphic and filamentous fungi: Cellular location of reduced selenium and implications for tolerance. J Ind Microbiol 14:300–311.

Harrison, J.J., Ceri, H., Stremick, C., Turner, R.J. (2004) Differences in biofilm and planktonic cell mediated reduction of metalloid oxyanions. FEMS Microbiol Lett 235:357–362.

Hou, J., Miao, L., Wang, C., Wang, P., Ao, Y., Qian, J., Dai, S. (2014) Inhibitory effects of ZnO nanoparticles on aerobic wastewater biofilms from oxygen concentration profiles determined by microelectrodes. J Hazard Mater 276:164–170.

Kucharíková, S., Tournu, H., Lagrou, K., van Dijck, P., Bujdáková, H. (2011) Detailed comparison of *Candida albicans* and *Candida glabrata* biofilms under different conditions and its susceptibility to caspofungin and anidulafungin. J Med Microbiol 60:1261–1269.

Lauchnor, E.G., Semprini, L. (2013) Inhibition of phenol on the rates of ammonia oxidation by *Nitrosomonas europaea* grown under batch, continuous fed, and biofilm conditions. Water Res 47:4692–4700.

Lemly, A.D. (2004) Aquatic selenium pollution is a global environmental safety issue. Ecotoxicol Environ Safety 59:44–56.

Lenz, M., Lens, P.N.L. (2009) The essential toxin: The changing perception of selenium in environmental sciences. Sci Total Environ 407:3620–3633.

Lewandowski, Z., Walser, G., Characklis, W. (1991) Reaction kinetics in biofilms. Biotechnol Bioeng 38:877–882.

Lorenzen, J., Glud, R.N., Revsbech, N.P. (1995) Impact of microsensor-caused changes in diffusive boundary layer thickness on O_2 profiles and photosynthetic rates in benthic communities of microorganisms. Mar Ecol Prog Ser 119:237–241.

Mah, T.F.C., O'Toole, G.A. (2001) Mechanisms of biofilm resistance to antimicrobial agents. Trends Microbiol 9:34–39.

Miller, G.L. (1959) Use of dinitrosalicylic acid reagent for determination of reducing sugar. Anal Chem 31:426–428.

Nancharaiah, Y.V., Lens, P.N.L. (2015) Selenium biomineralization for biotechnological applications. Trends Biotechnol 33:323–330.

Navarrete, F., De La Fuente, L. (2014) Response of *Xylella fastidiosa* to zinc: decreased culturability, increased exopolysaccharide production, and formation of resilient biofilms under flow conditions. Appl Environ Microbiol 80:1097–1107.

Papagianni, M. 2004. Fungal morphology and metabolite production in submerged mycelial processes. Biotechnol. Adv. 22, 189–259.

Pettit, R.K., Repp, K.K., Hazen, K.C. (2010) Temperature affects the susceptibility of *Cryptococcus neoformans* biofilms to antifungal agents. Med Mycol 48:421–426.

Ramage, G., Rajendran, R., Sherry, L., Williams, C. (2012) Fungal biofilm resistance. Int J Microbiol 2012:1–14.

Ramsay, J.A., Cooper, D.G., Neufeld, R.J. (1989) Effects of oil reservoir conditions on the production of water-insoluble levan by *Bacillus licheniformis*. Geomicrobiol. 7:155–165.

Sarkar, J., Dey, P., Saha, S., Acharya, K. (2011) Mycosynthesis of selenium nanoparticles. Micro Nano Lett 6:599–602.

Templeton, A.S., Trainor, T.P., Spormann, A.M., Brown, G.E. (2003) Selenium speciation and partitioning within *Burkholderia cepacia* biofilms formed on α-Al_2O_3 surfaces. Geochimica Cosmochimica Acta 67:3547–3557.

Tien, M., Kirk, T.K. (1988) Lignin peroxidase of *Phanerochaete chrysosporium*. In: Wood, W.A., Kellogg, S.T. (Eds.), Methods in Enzymology, Biomass, Part B: Lignin, Pectin and Chitin, vol. 161. Academic, San Diego, pp. 238-249.

USHHS (2003) Toxicological profile for selenium. Toxicological Profile for Selenium. Available at: http://www.atsdr.cdc.gov/toxprofiles/tp.asp?id=153&tid=28 Accessed 10 November 2015.

van Hullebusch, E.D., Zandvoort, M.H., Lens, P.N.L. (2003) Metal immobilization by biofilms: mechanisms and analytical tools. Rev Environ Sci Biotech 2:9–33.

Vetchinkina, E., Loshchinina, E., Kursky, V., Nikitina, V. (2013) Reduction of organic and inorganic selenium compounds by the edible medicinal basidiomycete *Lentinula edodes* and the accumulation of elemental selenium nanoparticles in its mycelium. J Microbiol 51:829–835.

Villa, F., Pitts, B., Stewart, P.S., Giussani, B., Roncoroni, S., Albanese, D., Giordano, C., Tunesi, M., Cappitelli, F. (2011) Efficacy of zosteric acid sodium salt on the yeast biofilm model *Candida albicans*. Microb Ecol 62:584–598.

Walters III, M.C., Roe, F., Bugnicourt, A., Franklin, M.J., Stewart, P.S. (2003) Contributions of antibiotic penetration, oxygen limitation, and metabolic activity to tolerance of *Pseudomonas aeruginosa* biofilms to ciprofloxacin and tobramycin. Microb Agents Chemother 47:317–323.

Wanner, O., Eberl, H.J., Morgenroth, E., Noguera, D., Picioreanu, C., Rittmann, B.E., van Loosdrecht, M. (2006) Mathematical modeling of biofilms. IWA Publishing, Hove, UK.

Wäsche S., Horn H., Hempel D.C. (2002) Influence of growth conditions on biofilm development and mass transfer at the bulk/biofilm interface. Water Res 36:4775–4784.

Yang, S.I., George, G.N., Lawrence, J.R., Kaminskyj, S.G.W., Dynes, J.J., Lai, B., Pickering, I.J. (2016) Multispecies biofilms transform selenium oxyanions into elemental selenium particles: studies using combined synchrotron X-ray fluorescence imaging and scanning transmission X-ray microscopy. Environ Sci Technol, Just accepted manuscript, DOI: 10.1021/acs.est.5b04529.

Yu, F.P., Callis, G.M., Stewart, P.S., Griebe, T., Mcfeters, G.A. (1994) Cryosectioning of biofilms for microscopic examination. Biofuling 8:85–91.

Zhang, T.C., Bishop, P.L. (1994) Evaluation of tortuosity factors and effective diffusivities in biofilms. Wat Res 28:2279–2287.

Zhang, X., Yan, S., Tyagi, R.D., Surampalli, R.Y. (2011) Synthesis of nanoparticles by microorganisms and their application in enhancing microbiological reaction rates. Chemosphere 82:489–494.

Zhou, X.H., Tong, Y., Han-Chang, S., Hui-Ming, S. (2011) Temporal and spatial inhibitory effects of zinc and copper on wastewater biofilms from oxygen concentration profiles determined by microelectrodes. Wat Res 45:953–959.

CHAPTER 7

Biomineralization of tellurium and selenium-tellurium nanoparticles by the white-rot fungus *Phanerochaete chrysosporium*

A modified version of this chapter was submitted as:

E.J. Espinosa-Ortiz, E.R. Rene, F. Guyot, E.D. van Hullebusch, P.N.L. Lens. (2016) Biomineralization of tellurium and selenium-tellurium nanoparticles by the white-rot fungus *Phanerochaete chrysosporium*.

Abstract

The release of tellurium and selenium, elements belonging to the chalcogen's group, in the environment can cause severe consequences to living organisms because of their toxicity. The white-rot fungus *Phanerochaete chrysosporium* was investigated for its response against tellurite (TeO_3^{2-}, 10 mg Te L^{-1}) and the concurrent exposure to selenite and tellurite (Se:Te, 10 mg Se L^{-1} + 10 mg Te L^{-1}) in batch conditions (pH 4.5, 30 °C, 150 rpm). The response of *P. chrysosporium* to selenite (SeO_3^{2-}, 10 mg Se L^{-1}) was also assessed and used as a reference since the SeO_3^{2-} inhibitory effects and influence on fungal morphology have already been described (Espinosa-Ortiz et al., 2015). *P. chrysosporium* was inhibited (fungal growth and substrate consumption) according to the corresponding incubation: Se:Te > SeO_3^{2-} > TeO_3^{2-}. Fungal morphology was influenced by the presence of both TeO_3^{2-} and SeO_3^{2-}. *P. chrysosporium* was identified as a Te-reducing organism capable to synthesize crystalline hexagonal elemental Te needle-like particles (20 to 465 nm). Se:Te incubations resulted in the biomineralization of both spherical and needle-like particles in the nano-size range (50-600 nm) comprising of both Te^0 and Se^0, which were intimately colocalized, although the Se:Te ratio was different in the two morphotypes. The ability of *P. chrysosporium* as a Se- and Te-reducing organism opens up the possibility to exploit this fungus for biotechnology and bioremediation applications, as well as for the biosynthesis of unique nanoparticles.

Keywords: *Phanerochaete chrysosporium*, biomineralization, tellurite, selenite, nanoparticles

7.1 Introduction

Tellurium (Te), a metalloid belonging to the chalcogens group (VI-A group of the periodic table), is scarcely and unevenly distributed in the earth's crust. Te oxyanions can be found in agricultural or industrial lands (Perkins, 2011), mine tailings or industrial effluents, but also in natural waters such as rivers and sea water (Biver et al., 2015). This element is of particular interest in virtue of its thermal, optical and semiconducting properties, with a broad range of applications, including alloying agent, electronics, rechargeable batteries, thermoelectric materials, cooling devices and solar panels (Turner et al., 2012). Reviews on the uniqueness of this metalloid, its biological potential and its microbial processing as a tool in biotechnology show the growing interest towards Te (Ba et al., 2010; Chasteen et al., 2009; Chivers and Laitinen, 2015; Turner et al., 2012). Due to its scarcity in nature, the exploration for new sources of Te or its recovery is imperative.

Te is not an essential trace element, but its toxicity towards most microorganisms has been documented; for some bacteria even at concentrations as low as 1 mg L^{-1} (Taylor, 1999; Díaz-Vásquez et al., 2015; Reinoso et al., 2012). The release of the toxic water soluble oxyanions of Te, tellurite (TeO_3^{2-}) and tellurate (TeO_4^{2-}), into the environment causes great concern; particularly by TeO_3^{2-}, which is considered as the most toxic one (Chasteen et al., 2009). Different microorganisms have been found capable to intra- or extracellularly reduce Te oxyanions into elemental Te (Te^0), which is less toxic and more stable. Bacteria have been the most studied Te-reducing organisms (Turner et al., 2012). Fungi have, however, also been suggested as potential Te-reducing organisms (Ba et al., 2010), capable of producing Te^0 as well as Te methylated compounds (Chasteen and Bentley, 2003).

Simultaneous existence of Te and selenium (Se), another chalcogen whose oxyanions can be highly toxic (Fordyce, 2007), occurs for example in copper and sulfur bearing ores, as well as in metal refinery wastewater (Soda et al., 2011). Although the effects of the concurrent exposure to Te and Se oxyanions on microorganisms (metabolical, physiological and morphological) have not yet been fully understood and exploited, the interaction among the two metalloids seems to have unique effects on their biomineralization. The presence of Se, as selenite (SeO_3^{2-}), triggers TeO_3^{2-} reduction in heterotrophic aerobic bacteria, accelerating the Te removal rate (Bajaj and Winter, 2014). For the first time, Bajaj and Winter (2014) reported the bacterial formation of extracellular nanospheres composed of both Te and Se. The simultaneous bioreduction of Te and Se oxyanions is promising in view of creating alternative biotechnologies for the removal of both metalloids from polluted effluents as well as for the green synthesis of unique biocomposites.

In this study, the Se-reducing white rot-fungus *Phanerochaete chrysosporium* (Espinosa-Ortiz et al., 2015a) was investigated for its response upon exposure to TeO_3^{2-} (10 mg Te L^{-1}) or exposure to TeO_3^{2-} and SeO_3^{2-} concurrently (10 mg Te L^{-1} + 10 mg Se L^{-1}). Effects on the fungal growth, activity and morphology were determined, as well as the potential of the fungus to remove Te and Se from polluted effluents. Different Se-Te ratios (1:1, 2:1, 4:1 and 1:2) were used to elucidate the effect of the presence of Se on TeO_3^{2-} reduction. Transmission Electron Microscopy (TEM) was used to characterize the biocomposites formed during oxyanion reduction.

7.2 Materials and methods

7.2.1 Fungal strain and culturing conditions

The fungal strain *P. chrysosporium* MTCC 787 (Institute of Microbial Technology, Chandigarh, India) was grown at 37 °C in malt extract agar plates. After 3 days of incubation, the fungal spores were harvested and used to prepare a fungal spore solution. Fungal incubations were prepared in 100 mL Erlenmeyer flaks with 50 mL of liquid medium. In a typical procedure, 10 g of glucose, 2 g of KH_2PO_4, 0.5 g of $MgSO_4{\cdot}7H_2O$, 0.1 g of NH_4Cl, 0.1 g of $CaCl_2{\cdot}2H_2O$, 0.001 g of thiamine, and 5 mL of trace element solution (Tien and Kirk, 1988) were dissolved in 1 L of deionized water and sterilized at 123 kPa and 110 °C for 30 min and cooled at room temperature before use. The pH was adjusted to 4.5. The spore solution was used as inoculum (2% *v/v*) for pre-cultivation cultures and incubated at 30 °C for 2 days. Flasks were placed in an orbital shaker (Innova 2100, New Brunswick Scientific) set at 150 rpm. The final inoculum consisted of sub-cultures of the 2 day-old fungus (2% *v/v*) as described previously (Espinosa-Ortiz et al., 2015a).

7.2.2 Batch experiments

7.2.2.1 Exposure to Se and Te oxyanions

The response of *P. chrysosporium* to the presence of SeO_3^{2-}, TeO_3^{2-} and their combination (Se:Te) was assessed. *P. chrysosporium* was incubated for 8 days in the presence of Na_2SeO_3 (10 mg Se L^{-1}), K_2TeO_3 (10 mg Te L^{-1}) or their combination (Se:Te, 10 mg Se L^{-1} and 10 mg Te L^{-1}). Incubations with SeO_3^{2-} were used as a reference of the response of *P. chrysosporium*, as it has already been demonstrated that SeO_3^{2-} decreases the biomass production of *P. chrysosporium* and induces overt morphological changes to the fungus (Espinosa-Ortiz et al., 2015a).

Evolution of the fungal growth and removal of the oxyanions was determined over time, sacrificing samples after each sampling step for analysis at 0, 1, 2, 3, 4, 6 and 8 days. Biotic controls (untreated) in the absence of any toxic oxyanion were performed as well. Abiotic controls (no biomass added) were also performed to determine abiotic reactions. All incubations and sacrificing samples were performed in triplicates.

7.2.2.2 Incubation with simultaneous presence of TeO_3^{2-} and SeO_3^{2-}

The effect of the SeO_3^{2-} concentration on the TeO_3^{2-} reduction was investigated by incubating *P. chrysosporium* with a constant initial TeO_3^{2-} concentration (10 mg Te L^{-1}) and with 5 to 40 mg Se L^{-1}, representing Se:Te ratios of 1:2, 1:1, 2:1 and 4:1, respectively. Experimental conditions were similar to those described above. Incubations in this experiment were maintained for 4 days, which was observed (from the time-dependent study described above) to be the period at which most fungal biomass is produced. Triplicates were conducted for each incubation.

7.2.3 Characterizations of fungal morphology

Macro-morphology of the fungal pellets was determined through digital images (Canon EOS Rebel T3, Taiwan). To find out if any changes of coloration occurred inside of the fungal pellets, samples were taken from the Erlenmeyer flasks after being incubated for 8 days and transferred manually into Petri dishes. Then the pellets were dissected with the help of a blade.

The micro-morphology of pellets was determined with TEM images. Samples were fixed in a 2.5% glutaraldehyde phosphate-buffered saline solution (pH 7.0), and serially dehydrated with 50%, 60%, 70%, 80%, 90% and 100% ethanol (Ngwenya and Whiteley, 2006). Samples were placed on a carbon-coated copper grid and imaged. TEM images and selected area electron diffraction (SAED) patterns were obtained with a JEM-2100F field emission electron microscope (JEOL, USA) equipped with a field effect gun operating at 200 kV. The distribution map of Se and Te in the samples was determined by Energy Dispersive X-ray Spectroscopy (EDS) analysis performed in Scanning TEM (STEM) mode. Images and diffraction patterns were processed using Image-J software (version 1.47, National Institute of Health, USA) (Rasband 1997-2014).

7.2.4 Analytical methods

After the corresponding incubation period, samples were centrifuged at 37,000 × g for 15 min, and the supernatant was separated and used for analysis including determination of glucose and total soluble residual Se and Te concentration in the growth medium. The dinitrosalicylic acid method was used for the analysis of glucose in medium with D-glucose as standard (Miller, 1959). Samples were further filtered (0.45 μm) and preserved in 0.5% HNO_3 solution in ultrapure water (Milli-Q water, 19 MΩ cm) for the analysis of the total Se and Te content, which was measured with graphite furnace atomic absorption spectroscopy (Thermo Elemental, Solaar MQZE, USA), with 2 μg L^{-1} as detection limit. Measurement of the volatile Se or Te fractions by *P. chrysosporium* was not performed in this study.

Fungal growth was measured based on the total dry biomass produced, which was determined gravimetrically. The fungal suspension was filtered using a 0.45 μm-pore size filter paper (Type GF/F, Whatman Inc., Florham Park, NJ), which was previously dried (24 h at 105 °C) and weighted. The Se and Te removal rates were estimated as previously described for batch conditions by Espinosa-Ortiz et al. (2015b).

7.3 Results

7.3.1 Fungal interaction with chalcogen oxyanions

The response of *P. chrysosporium* to the presence of TeO_3^{2-} and its concurrent combination with SeO_3^{2-} was assessed in terms of morphology, fungal growth, substrate consumption and removal of Se and Te, after 8 days of incubation (Figure 7.1, Table 7.1). Compared to the untreated incubations, fungal growth decreased around 46 (± 3)% for the TeO_3^{2-} incubations, and around 83.9 (± 1.5)% for the SeO_3^{2-} and Se:Te incubations (Figure 1A). The biomass yield decreased according to the corresponding incubation: untreated > TeO_3^{2-} > SeO_3^{2-} > Se:Te (Table 7.1). The biomass yield was clearly decreased in incubations containing SeO_3^{2-} in comparison to incubations containing TeO_3^{2-}: with respect to the Se-free incubations, the biomass yield was only 60% and 40% for SeO_3^{2-} and Se:Te incubations, respectively, whereas incubations with TeO_3^{2-} had only 84% of the control biomass yield. Consumption of substrate, which is an indicator of the metabolic activity of *P. chrysosporium*, was also limited by the presence of the oxyanions: SeO_3^{2-} and Se:Te incubations showed a similar trend of inhibition of the glucose consumption (Figure 7.1B). The final pH of the incubations in the presence of the individual oxyanions or their combination decreased from 4.50 to 3.08 (± 0.02) in average.

Table 7.1 Fungal growth and morphology of *P. chrysosporium* when incubated with TeO_3^{2-}, SeO_3^{2-}, or a combination of both, incubated over a period of 8 d.

Incubation	Shape	Surface	Color	Dry weight (g L^{-1})[a]	$Y_{x/s}$ (g·g^{-1})[b]
Untreated	Pellet	Hairy	Ivory white	1.41±0.01	0.25±0.001
Tellurite	Pellet	Smooth	Black	0.76±0.06	0.21±0.03
Selenite	Pellet	Smooth	Red-orange	0.25±0.04	0.14±0.001
Selenite:Tellurite	Pellet	Smooth	Black	0.22±0.03	0.10±0.02

[a]Values are means (n=3) with standard deviations

[b]Biomass yield, calculated as $Y_{X/S} = \frac{g\ biomass\ produced\ (\Delta X)}{g\ substrate\ utilized\ (\Delta S)}$

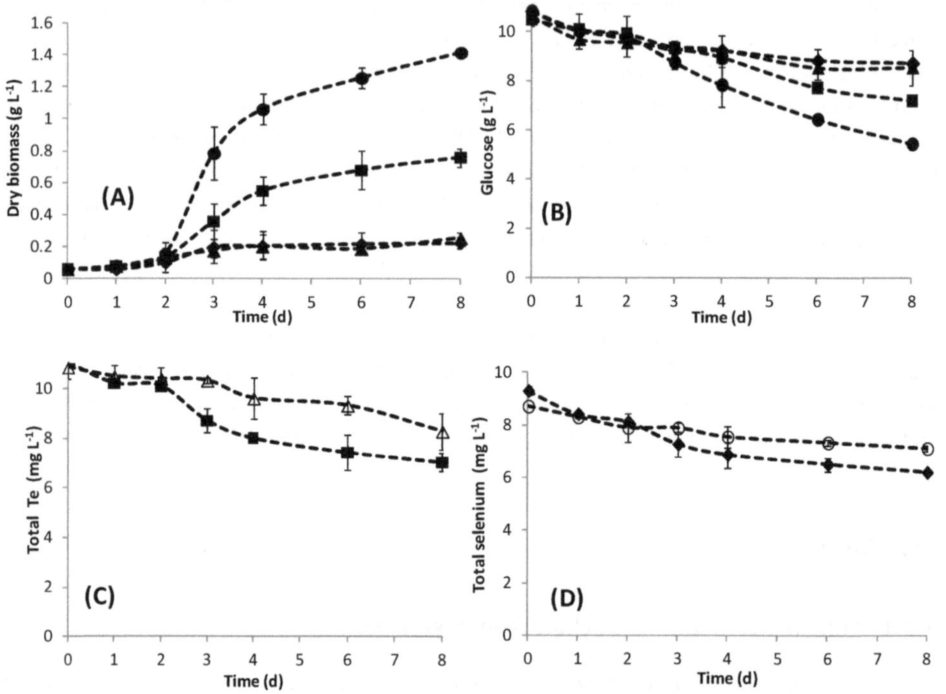

Figure 7.1 Response of *P. chrysosporium* to tellurite (10 mg Te L^{-1}), selenite (10 mg Se L^{-1}) and their combination (10 mg Te L^{-1}+10 mg Se L^{-1}) after 8 days of incubation (pH 4.5, 10 g glucose L^{-1}, 30 °C). (A) Fungal growth, (B) Substrate consumption, (C) Te removed and (D) Se removed. ● Untreated, ■ Tellurite, ◆ Selenite, ▲ Selenite:Tellurite (Se:Te), △ Se:Te-Te, ○ Se:Te-Se.

Total soluble residual concentrations of Te and Se were determined (Figures 7.1C and D). In average, Te removal in TeO_3^{2-} incubations was 36.3 (± 2.7)% (0.48 ± 0.02 mg Te $L^{-1}d^{-1}$) and 23.7 (± 3.6)% (0.32 ± 0.04 mg Te $L^{-1}d^{-1}$) in Se:Te incubations. Removal of Se was around 33.3 (± 1)% (0.39 ± 0.01 mg Se $L^{-1}d^{-1}$) in SeO_3^{2-} incubations and 18.3 (± 1.1)% (0.20 ± 0.01 mg Se $L^{-1}d^{-1}$) in Se:Te incubations. No changes in the initial concentrations of Se or Te were observed in the abiotic incubations (data not shown), which suggests that Se and Te removal is related to metabolic activities of *P. chrysosporium*. A characteristic garlic-like odor, detected while respecting all safety regulations, was perceived in all the incubations with Se (SeO_3^{2-}, Se:Te), particularly strong in SeO_3^{2-} incubations, suggesting the formation of volatile organic Se species. No odor was perceived in TeO_3^{2-} incubations.

A black coloration of the fungal pellets treated with TeO_3^{2-} (Figure 7.2) was indicative of the reduction of TeO_3^{2-} to Te^0. Se:Te incubations also presented a black coloration, which could be indicative of the presence of Te^0, although Se-Te alloys have also been characterized by a black color (Zhou and Zhu, 2006). No sign of red-orange coloration, indicative of the presence of Se^0, was observed in the fungal biomass of Se:Te incubations (Figure 7.2), even when pellets were dissected to search for evidence of coloration in the core of the fungal pellet.

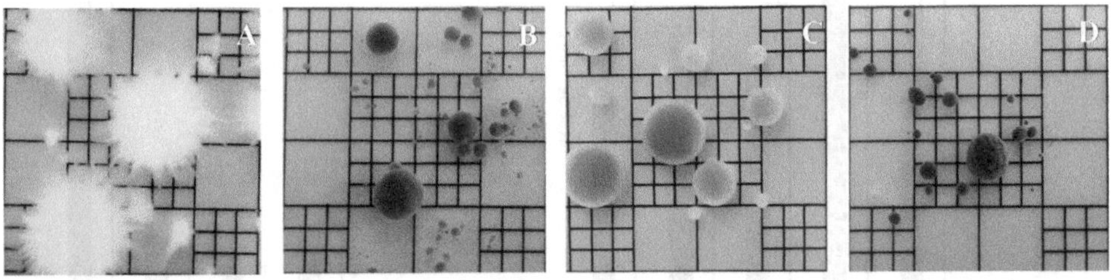

Figure 7.2 Macro-morphology of *P. chrysosporium* pellets after 8 days of incubation (pH 4.5, 10 g glucose L^{-1}, 30 °C). (A) Untreated, (B) Tellurite (10 mg Te L^{-1}), (C) Selenite (10 mg Se L^{-1}) and (D) Selenite:Tellurite (10 mg Te L^{-1}+10 mg Se L^{-1}).

7.3.2 Effect of Se:Te ratio on TeO_3^{2-} reduction

The response of *P. chrysosporium* to the presence of both TeO_3^{2-} and SeO_3^{2-} varying the Se:Te ratio (0:1, 1:1, 1:2, 2:1 and 4:1) was investigated after 4 days of incubation (Figure 7.3). Compared to the untreated incubations, biomass growth decreased ~45% in the presence of TeO_3^{2-} (Se:Te, 0:1), whereas a 70% decrease was observed for Se-Te incubations (Figure 7.3A). Fungal growth inhibition depended on the initial SeO_3^{2-} concentration; the higher the SeO_3^{2-} concentration the higher the growth inhibition, limiting the growth up to almost 90% (Se:Te, 4:1). The metabolic activity of the fungus was also influenced by the presence of SeO_3^{2-} in the medium, showing that the substrate consumption was limited by increasing SeO_3^{2-} concentrations (Figure 7.3B).

Maximal Te removal was observed in the absence of SeO_3^{2-} (ratio 0:1, 32.4 ± 5%). The incubations in the presence of Se:Te with a 1:1 ratio showed a Te removal efficiency of 10.5 (± 2.4)%, whereas the rest of the incubations (ratios 1:2, 2:1 and 4:1) showed less than 10% Te removal (Figure 7.3C). The treatment with a 1:2 Se:Te ratio (5 mg Se L^{-1} and 10 mg Te L^{-1}) gave the lowest Te removal. The removal of total Se was also determined. At equal mass concentrations of SeO_3^{2-} and TeO_3^{2-} (ratio 1:1), total Se removal reached 16.6 (± 0.5)%. With half the concentration of TeO_3^{2-} (Se:Te ratio of 2:1),

Se was removed ~20%. At high Se concentrations (ratios 2:1 and 4:1), removal of SeO_3^{2-} was less than 8% (Figure 7.3D).

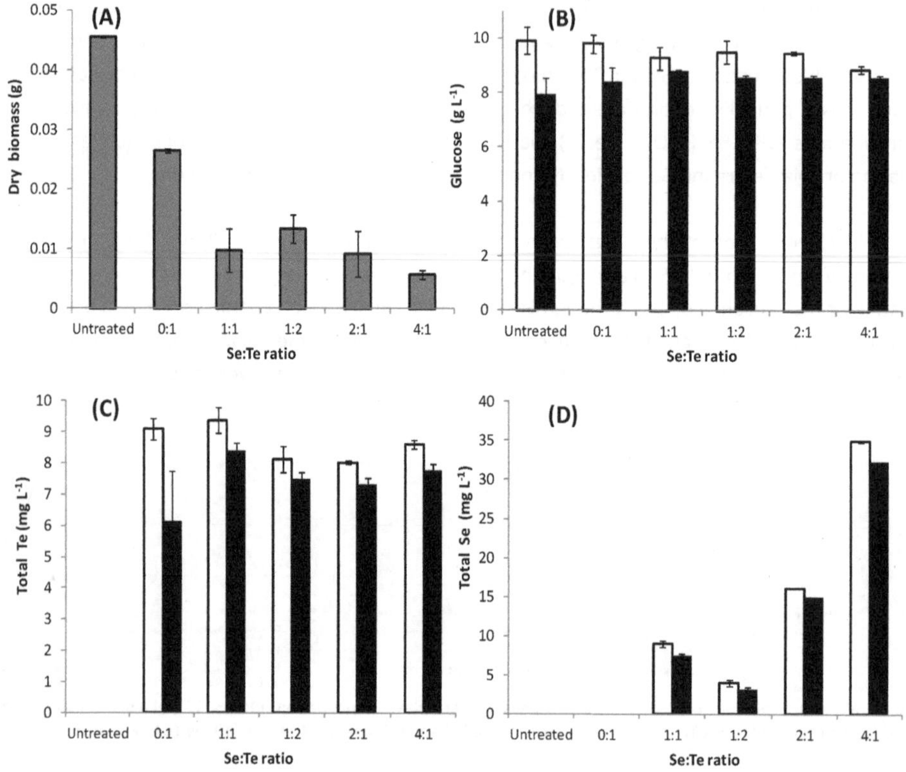

Figure 7.3 Effects on *P. chrysosporium* after 4 days of incubation (pH 4.5, 10 g glucose L^{-1}, 30 °C) when selenite and tellurite are present concurrently at different ratios. (A) Fungal growth, (B) Substrate consumption, (C) Tellurium removed and (D) Se removed.
□ t=0 days, ■t=4 days.

7.3.3 Fungal morphology

Table 7.1 and Figures 7.2A, B, C and D show the macro-morphology of the untreated, TeO_3^{2-}, SeO_3^{2-} and Se:Te treated incubations, respectively. Regardless of the presence or absence of Se and Te in the growth medium, *P. chrysosporium* grew as pellets. As previously described (Espinosa-Ortiz et al., 2015a), untreated fungi grew as ivory-white pellets with fibril-like structures (Figure 7.2A), whereas fungi incubated with SeO_3^{2-} formed more compact and smoother pellets almost as perfect spheres (Figure 7.2B). Incubations with TeO_3^{2-} produced compact and smooth pellets ranging from perfect spheres to irregular forms (elongated pellets) (Figure 7.2C). Almost perfect smooth and compact spheres were also obtained when *P. chrysosporium* was incubated concurrently with Se and Te (Figure 7.2D).

7.3.4 Electron microscopic analysis

Fungal pellets exposed to TeO_3^{2-} (10 mg Te L^{-1}), SeO_3^{2-} (10 mg Se L^{-1}) and Se:Te (1:1 mass ratio) for 4 days of incubation were used for TEM analysis. TEM images showed the formation of Te^0 in the fungal hyphae when incubated with TeO_3^{2-} (Figure 4A). Needle-like particles of Te^0 (Figure 7.4B) in the nano-size range (20-465 nm) were observed. TEM images revealed the accumulation of these particles in clusters (Figures 7.4D and E). STEM-EDS analysis confirmed that the particles were entirely comprised of Te (data not shown). The SAED pattern (Figure 7.4C) indicated that the nanoparticles were well crystallized corresponding to hexagonal Te (Keller et al., 1977). SeO_3^{2-} led to the formation of amorphous nanospheres (Figure 7.4F) as described previously (Espinosa-Ortiz et al., 2015a).

P. chrysosporium incubated in the presence of SeO_3^{2-} and TeO_3^{2-} (1:1 mass ratio) biomineralized Se-Te nanoparticles (Figure 7.5A). Two morphotypes were observed: i) needle like particles (Figures 7.5B and D) ranging from 50 to 600 nm in size, and ii) spheres (Figure 7.5D) in the range of 70 to 300 nm in size. The SAED pattern (Figure 7.5C) indicated that the needle-like nanoparticles were crystalline, very similar to the corresponding hexagonal Te nanoparticles (Keller et al., 1977), consistent with unit cell parameters a:4.280(1) Å and c:5.967(2) Å, space group 152, JCPDS card 01-085-0558. An amorphous structure was determined for the Se-Te (Se-rich) nanospheres (data not shown). The STEM-EDS analysis confirmed an intimate co-localization of Se and Te in both morphotypes (Figures 7.5E and F). Elemental analysis in multiple sites of each single structure was performed (data not shown), which confirmed the presence of both Se and Te in each structure but in different proportions. Qualitatively, the spheres had a higher Se content (Se-Te, 7:3), whereas the needle-like structures possessed a higher Te content (Se-Te, 4:6).

Abiotic controls were performed (incubations in the absence of fungi) to account for possible formation of Te or Se precipitates due to the operational conditions in the liquid medium during the TeO_3^{2-}, SeO_3^{2-} and Se:Te incubations. However, no Te or Se precipitates were observed in any of the incubations (data not shown).

7.4 Discussion

7.4.1 *P. chrysosporium* as a TeO_3^{2-}-reducing organism

This study investigated for the first time the response of the white-rot fungus *P. chrysosporium* to the presence of TeO_3^{2-} and found it to be capable of reducing TeO_3^{2-} to Te^0. Although the biochemical mechanism used to reduce TeO_3^{2-} is not clear yet, it is assumed that the fungus mimics Se detoxification mechanisms (Ba et al., 2010). Reductive processes are involved in the microbial tolerance for and resistance to TeO_3^{2-}, leading to the synthesis of Te^0 or the formation of volatile species through a series of different mechanisms including enzymatic reduction, reduction with thiols (*e.g.* glutathione) and methylation (Turner et al., 2012; Zannoni et al., 2008).

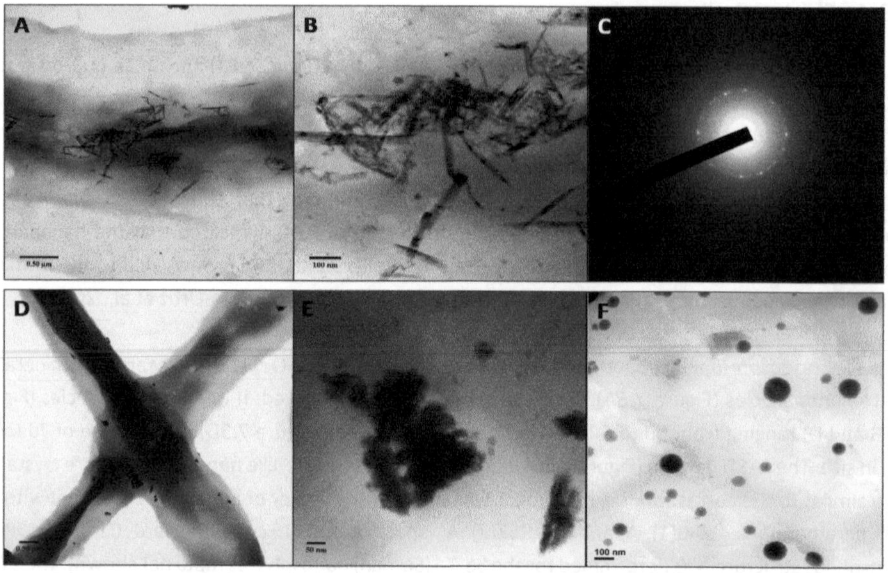

Figure 7.4 TEM images of Te and Se nanoparticles biomineralized by *P. chrysosporium* after 4 days of incubation (pH 4.5, 10 g glucose L^{-1}, 30 °C, 10 mg Te L^{-1} or 10 mg Se L^{-1}). (A) Needle-like Te^0 particles in fungal hyphae, (B) Close up of the needle-like Te^0 particles, (C) SAED pattern for needle-like Te^0 particles (main diffraction rings measured at around 3.3 Å and 2.4 Å), (D) Distribution of Te^0 particles in fungal hyphae, Close up of the (E) Te^0 particle aggregates and (F) Se^0 nanospheres.

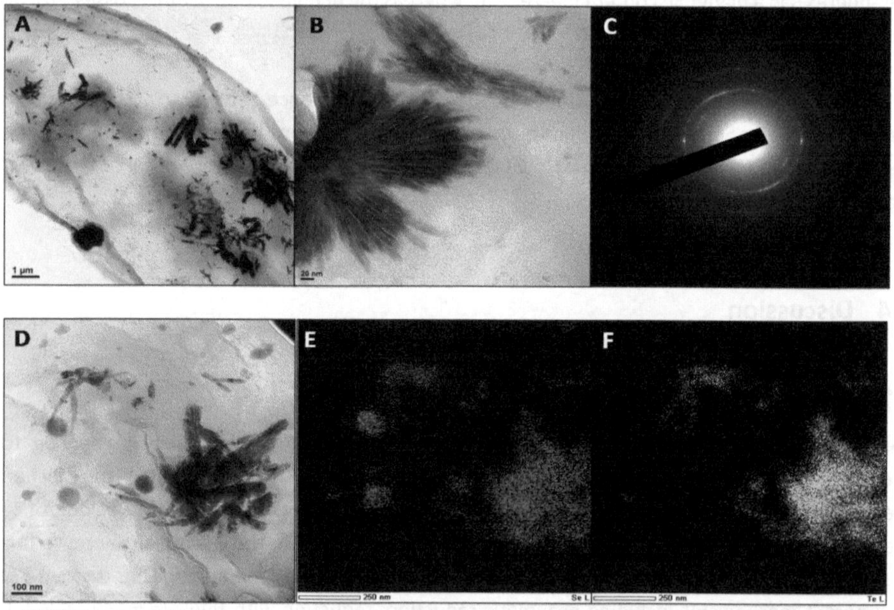

Figure 7.5 TEM images of Se-Te nanoparticles synthesized by *P. chrysosporium* after 4 days of incubation (pH 4.5, 10 g glucose L^{-1}, 30 °C, 10 mg Te L^{-1}+10 mg Se L^{-1}). (A) Distribution of particles of different sizes within fungal biomass, (B) Close up to Se-Te particles, (C) SAED pattern of needle-like Se-Te nanoparticles (main diffraction rings measured at around 4.0 Å, 3.3 Å, 2.35 Å and 1.95 Å), (D) Different particles in the hyphae and STEM-EDS elemental mapping for (E) Se and (F) Te. It should be noted that Te is present in the Se-Te spheres although it is hardly visible in the elemental mapping.

A black coloration of the fungal biomass was observed when treated with TeO_3^{2-} (Figure 7.2B), which is a manifestation of the reduction of TeO_3^{2-} to Te^0, as previously reported for bacteria (Borghese et al., 2014; Kim et al., 2012). This finding is in agreement with previously reported work with other fungal strains. Gharieb et al. (1999) observed intra- and extracellular black deposits of Te^0 in *Fusarium* when exposed to 50 mM (~8.7 g L^{-1}) of TeO_3^{2-}. The authors suggested the deposition of Te^0 in fungal vacuoles. Vacuoles play a key role in the accumulation, compartmentation and detoxification of Se and Te, as demonstrated with the yeast *Saccharomyces cerevisiae*, which accumulated Se and Te in the cellular cytosolic compartment (Gharieb and Gadd, 1998).

The absence or insignificant amount of volatile products in the TeO_3^{2-} incubations is suggested by the absence of a garlic-like odor, which is associated to the production of dimethyl telluride, the most common volatile biomethylated product of Te (Chasteen and Bentley, 2003). Gharieb et al. (1999) suggested that methylation of Te species does not play a major role as a Te detoxification mechanism in fungi. Further studies on the capture of volatile Te species are required to confirm this finding.

This study shows *P. chrysosporium* is promising as a potential agent to remove TeO_3^{2-} from effluents, particularly for the treatment of acidic effluents. This fungus was able to grow and form pellets in the presence of TeO_3^{2-}, with Te removal efficiencies of ~ 40% under mild acidic conditions (pH 4.5–3.1). Compared to bacteria, *P. chrysosporium* is not as efficient to remove TeO_3^{2-}. Most bacteria studied for the removal of TeO_3^{2-} reached more than 90% Te removal efficiency (initial Te concentration ranged between 100 µM to 1 mM, 12.7 to 127 mg Te L^{-1}) within a couple of days or even hours (Amoozegar et al., 2008; Amoozegar et al., 2012; Kim et al., 2012). Nevertheless, these optimal removal efficiencies were always obtained at neutral to basic pH conditions.

TeO_3^{2-} had a clear inhibitory effect on the growth of *P. chrysosporium* by decreasing the biomass production by about 40% compared to the untreated incubations. The influence of TeO_3^{2-} has been observed to be fungal species dependent (Gharieb et al. 1999; Prange et al. 2005). Incubated under the same conditions, *Penicillium citrinum* did not show a significant decrease in growth (biomass produced) in the presence of 1 mM Na_2TeO_3 (127 mg Te L^{-1}), whereas *Fusarium* showed a marked decrease in its biomass production (Gharieb et al., 1999). Incubations with SeO_3^{2-} strongly decreased fungal growth (~70%), as previously observed (Espinosa-Ortiz et al., 2015a). This clearly indicates that SeO_3^{2-} has a more inhibitory effect on the growth of *P. chrysosporium* than TeO_3^{2-}. However, the effect of SeO_3^{2-} compared to TeO_3^{2-} on the fungal growth cannot be generalized. Two different strains of *Aspergillus* showed to be less inhibited, in terms of biomass production, by SeO_3^{2-} compared to TeO_3^{2-} (Prange et al., 2005), whereas *Phomopsis viticola* showed the same degree of inhibition when incubated in the presence of SeO_3^{2-} or TeO_3^{2-} (Prange et al., 2005).

7.4.2 Synergetic effect of SeO_3^{2-} and TeO_3^{2-} on *P. chrysosporium*

Despite *P. chrysosporium* is a Se- and Te-reducing organism, a negative synergetic effect of SeO_3^{2-} and TeO_3^{2-} induced a noticeable decrease on the fungal growth compared to the untreated and TeO_3^{2-} incubations. In fact, the biomass production in Se:Te incubations was as low as in the incubations with SeO_3^{2-} (Figure 7.1A). Similarly, the metabolic activity of the fungus, represented by the fungal ability to consume glucose, decreased with the same tendency as that of the SeO_3^{2-} incubations (Figure 7.1B). The effect of different Se:Te ratios showed that fungal growth was the lowest in the incubations with the highest SeO_3^{2-} concentrations (Figure 7.3A). This highlights the stronger effect of SeO_3^{2-} compared to the effect induced by TeO_3^{2-} to the fungus, suggesting that *P. chrysosporium* is more inhibited by SeO_3^{2-} than by TeO_3^{2-}, on a mass concentration basis.

The reduction of TeO_3^{2-} was highly influenced by the presence of SeO_3^{2-}: the total Te removal efficiency was decreased to ≤10% for Se:Te incubations at different ratios (Figure 7.3). Bajaj and Winter (2014) found that TeO_3^{2-} reduction rates by Gram negative proteobacteria strains decreased according to decreasing SeO_3^{2-} concentrations. At a 1:1 ratio (100 mg Se/Te L^{-1}), no reduction of Se or Te occurred. However, the authors found that Te reduction was faster when SeO_3^{2-} was added in excess: the higher the SeO_3^{2-} concentration, the faster the reduction of TeO_3^{2-}. Opposite effects were observed in the present study: when *P. chrysosporium* was incubated at different Se:Te ratios, a faster reduction of TeO_3^{2-} due to SeO_3^{2-} was not observed. The removal of TeO_3^{2-} was inhibited in the presence of SeO_3^{2-}, reaching lower removal efficiencies during the same period compared to incubations without SeO_3^{2-}. SeO_3^{2-} does thus not induce a fast reduction of TeO_3^{2-} in *P. chrysosporium*, as observed for bacteria (Bajaj and Winter, 2014).

In the presence of both SeO_3^{2-} and TeO_3^{2-}, a light garlic-like odor was perceived, suggesting that methylated forms of either Se or Te are being produced. Fleming and Alexander (1972) found that *Penicillium* was able to produce dimethyltelluride in the presence of both Se and Te, characterized by a strong-garlic-like odor. In the absence of Se, *Penicillium* did not generate Te volatilization, suggesting that Se was involved in the activation of the methylation process in *Penicillium*, allowing then the formation of Te methylated forms (Fleming and Alexander, 1972).

7.4.3 Morphological effects induced by TeO_3^{2-} and Se-Te combinations

Fungal morphology is affected by several factors, including cultivation conditions, medium composition and fungal strain (Espinosa-Ortiz et al., 2016). Some filamentous fungi are able to self-immobilize in the form of pellets when incubated in liquid conditions as a response to limiting conditions, *e.g.* limited nutrients in the growth medium (Papagianni, 2004) or the presence of toxic species in the cultivation medium (Espinosa-Ortiz et al., 2015a; Saraswathy and Hallberg, 2005). Incubations with TeO_3^{2-} showed a less defined shape and smaller size in the fungal pellets compared to SeO_3^{2-} incubations. Se:Te incubations showed perfect spherical pellets. Bigger pellets represent better settleability, and therefore a more efficient separation of the fungal biomass from the treated effluent, which in turn can also lead to an easier separation and recovery of the synthesized nanoparticles. However, the diameter of the pellets should not exceed a critical size as this could lead to an insufficient supply of oxygen and nutrients within the fungal pellet (Wittier et al., 1986), restraining the ability of the fungus to reduce the metalloids and eventually leading to disintegration of the pellets.

7.4.4 Production of nTe^0 and nSe-Te

When grown with TeO_3^{2-}, *P. chrysosporium* synthesized nanoprecipitates consisting of needle-like nanoparticles of Te^0 having a hexagonal crystalline structure (Figures 7.4A and B). The size of the Te needle-like nanoparticles varied significantly (20-465 nm). These Te nanoparticles presumably grew from cellular nucleation sites, regions where TeO_3^{2-} reduction takes place. Formation of clusters of a bigger size was also observed (Figures 7.4D and E). The formation of clusters might be interpreted as the end product of TeO_3^{2-} reduction, based on their evident abundance in the fungal hyphae. The synthesis of Te^0 precipitates has also been observed in other fungal strains, however, the nanomaterials were not characterized. The microbial synthesis of Te^0 particles has been reported both intra- and extracellularly, mostly synthesized in the nano-size range in a broad range of different morphologies, including spheres, rods and needle-like nanoprecipitates (Table 7.2). Whereas similar external morphologies in Se^0 nanoparticles do not share comparable spectral properties (Oremland et al., 2004), different Te^0 biocomposites have showed similar internal structures, mainly attributed to the basic trigonal alignment features of the Te chains (Baesman et al., 2007).

This study shows that *P. chrysosporium* is able to biomineralize co-localized Se-Te nanoprecipitates of similar crystalline structure of that of the hexagonal Te (Figures 7.4 and 5). The biosynthesis of Se-Te nanoparticles has already been achieved with bacteria (Table 7.2), but to the best of our knowledge, the synthesis of Se-Te nanoparticles by other fungal species has not yet been reported. *P. chrysosporium* synthesized two different Se-Te morphotypes: nanospheres and needle-like nanoparticles (Figure 7.5) with different Se:Te proportions. The particles with higher proportion of Se showed similar morphology to that of the Se nanospheres (Figures 7.5D and 4F), whereas particles with higher proportion of Te had a similar structure to that of the Te^0 needle-like particles. Fleming and Alexander (1972) reported for the first time the concurrent interaction of SeO_3^{2-} and TeO_3^{2-} with *Fusarium*, but there was no indication of Se-Te precipitate formation.

It should be noted that this is the first time that needle-like Se-Te nanoprecipitates are biologically synthesized. Pearce et al. (2011) observed the formation of Se-Te nanoparticles under anaerobic conditions using *Bacillus beveridge*. TeO_3^{2-} (10 mM, 1.27 g Te L^{-1}) was first added to the bacterial suspension, which led to the formation of Te^0. Afterwards, SeO_3^{2-} (5 mM, 0.39 g L^{-1}) was added to the suspension, leading to its reduction to Se^0 and then to Se^{2-}. Although the synthesis mechanism of Se-Te nanoparticles was not clear, Pearce et al. (2011) suggested that such a reducing environment was created enabling the further reduction of Te^0 to Te^{2-}. Bajaj and Winter (2014) investigated the biogenic production of Se-Te nanoparticles by Se-reducing bacteria isolated from agricultural soils. The Se-Te nanospheres biosynthesis occurred when SeO_3^{2-} and TeO_3^{2-} were added concurrently to the growth medium, having similar initial oxidation states.

Se-Te materials are comprised of bound atoms arranged in helical chains, connected by van der Waals interactions forming anisotropic crystals (Berger, 1997). Se-Te nanoparticles have unique semiconductive and optical properties, as well as enhanced properties compared to single Te and Se nanomaterials, including higher electrical resistance and magnetoresistance (Sadtler et al., 2013; Sridharan et al., 2013), with potential uses in optoelectronics such as light emitting diode (LED) (Tripathi et al. 2009). Few studies have been dedicated to the chemical production of Se-Te nanomaterials (Fu et al., 2015; Qin et al., 2008; Zhou and Zhu, 2006). This study shows a promising green alternative approach for the synthesis of such Se-Te nanocomposites using *P. chrysosporium*. Further studies are required to characterize the properties of the Se-Te mycomineralized nanoparticles.

Table 7.2 Biomineralization of Te^0 and Se-Te composites by *P. chrysosporium*.

Microorganism	Nanomaterial	Reference

Bacillus selenitireducers	Te^0 nanorods, 10x200 nm, cluster ~1000 nm	Baesman et al., 2007
Bacillus sp.	Te^0 rod-like nanoparticles, 20x180 nm	Zare et al., 2012
Pseudomonas pseudoalcaligenes	Te^0 nanorods, 22x185 nm	Forootanfar et al., 2015
Rodhobacter capsulatus	Te^0 splinter-like nanoparticles, 80-300 nm	Borghese et al., 2014
Shewanella oneidensis	Te^0 nanorods, 100-200 nm	Kim et al., 2012
S. barnesii	Te^0 nanospheres, <50 nm	Baesman et al., 2007
Bacillus beveridgei	Se-Te nanospheres	Pearce et al., 2011
Non-halophilic aerobic bacteria	Se-Te nanospheres, 100 nm	Bajaj and Winter, 2014
P. chrysosporium	Te^0 needle-like nanoparticles, 20-465 nm Se-Te nanospheres and needle-like nanoparticles, 50-600 nm	This study

7.5 Conclusions

The ability of *P. chrysosporium* as a Te-reducing organism was demonstrated in this study. TeO_3^{2-} had an influence on the growth and morphology of the fungus. The use of *P. chrysosporium* to remove Te from contaminated mild acidic effluents is suggested. The recovery of Te^0 needle-like nanoparticles entrapped in the biomass is possible. The concurrent incubation with TeO_3^{2-} and SeO_3^{2-} had a higher inhibitory effect on the fungal growth, activity and morphology. Biomineralization of spheres and needle-like nanoparticles in the nano-size range comprised of both Te and Se. To the best of our knowledge, this is the first time that such biocomposites have been reported. The ability of *P. chrysosporium* as a Se- and Te-reducing organism opens up the possibility to exploit this fungus not only for biotechnology and bioremediation applications, but also for the synthesis of unique biocomposites.

7.6 References

Amoozegar M.A., Ashengroph M., Malekzadeh F., Reza Razavi M., Naddaf S., Kabiri M. (2008) Isolation and initial characterization of the tellurite reducing moderately halophilic bacterium, *Salinicoccus* sp. strain QW6. Microbiol Res 163(4):456–465.

Amoozegar M.A., Khoshnoodi M., Didari M., Hamedi J., Ventosa A., Baldwin S. (2012) Tellurite removal by a tellurium-tolerant halophilic bacterial strain, *Thermoactinomyces* sp. QS-2006. Annals Microbiol 62(3):1031–1037.

Anderson S. (2015) Mineral commodity summmaries - Selenium and tellurium. US Geological Survey.

Ba L.A., Döring M., Jamier V., Jacob C. (2010) Tellurium: an element with great biological potency and potential. Org Biomolecular Chem 8(19):4203–4216.

Baesman S.M., Bullen T.D., Dewald J., Zhang D., Curran S., Islam F.S., Oremland R.S. (2007) Formation of tellurium nanocrystals during anaerobic growth of bacteria that use Te oxyanions as respiratory electron acceptors. Appl Environ Microbiol 73(7):2135–2143.

Bajaj M., Winter J. (2014) Se (IV) triggers faster Te (IV) reduction by soil isolates of heterotrophic aerobic bacteria: formation of extracellular SeTe nanospheres. Microbial Cell Fact 13(1):1–10.

Berger L.I. (1997) Semiconductor Materials. Boca Raton, FL: CRC Press.

Biver M., Quentel F., Filella M. (2015) Direct determination of tellurium and its redox speciation at the low nanogram level in natural waters by catalytic cathodic stripping voltammetry. Talanta 144:1007–1013.

Borghese R., Baccolini C., Francia F., Sabatino P., Turner R.J., Zannoni D. (2014) Reduction of chalcogen oxyanions and generation of nanoprecipitates by the photosynthetic bacterium *Rhodobacter capsulatus*. J Hazard Mat 269:24–30.

Boriová K., Čerňanský S., Matúš P., Bujdoš M., Šimonovičová A. (2014) Bioaccumulation and biovolatilization of various elements using filamentous fungus *Scopulariopsis brevicaulis*. Lett Appl Microbiol 59(2):217–223.

Chasteen T.G., Bentley R. (2002) Biomethylation of selenium and tellurium: microorganisms and plants. American Chem Soc 103(1):1–25.

Chasteen T.G., Fuentes D.E., Tantaleán J.C., Vásquez C.C. (2009) Tellurite: history, oxidative stress, and molecular mechanisms of resistance. FEMS Microbiology Reviews, 33(4):820–832.

Chivers T., Laitinen R.S. (2015) Tellurium: a maverick among the chalcogens. Chem Soc Rev Chemical 44(c):1725–1739.

Díaz-Vásquez W.A., Abarca-Lagunas M.J., Cornejo F.A., Pinto C., Arenas F., Vásquez C.C. (2015) Tellurite-mediated damage to the *Escherichia coli* NDH-dehydrogenases and terminal oxidases in aerobic conditions. Archives Biochem Biophys 566:67–75.

Espinosa-Ortiz E.J., Gonzalez-Gil G., Saikaly P. E., van Hullebusch E. D., Lens P.N.L. (2015a) Effects of selenium oxyanions on the white-rot fungus *Phanerochaete chrysosporium*. Appl Microbiol Biotechnol 99(5):2405–2418.

Espinosa-Ortiz E.J., Rene E.R., van Hullebush E.D., Lens P.N.L. (2015b) Removal of selenite from wastewater in a *Phanerochaete chrysosporium* pellet based fungal bioreactor. Int Biodeterioration Biodegradation 102:361–269.

Espinosa-Ortiz E.J., Rene E.R., Pakshirajan K., van Hullebusch E.D., Lens P.N.L. (2016) Fungal pelleted reactors in wastewater treatment: applications and perspectives. Chem Eng J 283:553–571.

Fordyce F.F. (2007) Selenium geochemistry and health. Ambio 36:94–7.

Fu S., Cai K., Wu L., Han H. (2015) One-step synthesis of high-quality homogenous Te/Se alloy nanorods with various morphologies. Cryst Eng Comm 17:3243–3250.

Gharieb M.M., Gadd G.M. (1998) Evidence for the involvement of vacuolar activity in metal(loid) tolerance: Vacuolar-lacking and -defective mutants of *Saccharomyces cerevisiae* display higher sensitivity to chromate, tellurite and selenite. Bio Metals 11(2):101–106.

Gharieb M.M., Kierans M., Gadd G.M. (1999) Transformation and tolerance of tellurite by filamentous fungi: accumulation, reduction, and volatilization. Mycol Resear 103(3):299–305.

Keller R., Holzapfel W.B., Schulz H. (1977) Effect of pressure on the atom positions in Se and Te. Phys Rev B: Solid State 16:4404–4412.

Kim D.H., Kanaly R.A., Hur H.G. (2012) Biological accumulation of tellurium nanorod structures via reduction of tellurite by *Shewanella oneidensis* MR-1. Bioresour Technol 125:127–131.

Miller G.L. (1959) Use of dinitrosalicylic acid reagent for determination of reducing sugar. Anal Chem 31(3):426–428.

Ngwenya N., Whiteley C.G. (2006) Recovery of rhodium (III) from solutions and industrial wastewaters by a sulfate-reducing bacteria consortium. Biotechnol Prog 22:1604–1611.

Oremland R.S., Switzer B.J., Langley S., Beveridge T.J., Sutto T., Ajayan P.M., Curran, S. (2004) Structural and spectral features of selenium nanospheres formed by Se-respiring bacteria. Appl Environ Microbiol 70:52–60.

Papagianni M. (2004) Fungal morphology and metabolite production in submerged mycelial processes. Biotechnol Adv 22(3):189–259.

Pearce C.I., Baesman S.M., Blum J.S., Fellowes J.W., Oremland R.S. (2011) Nanoparticles formed from microbial oxyanion reduction of toxic group 15 and group 16 metalloids. In: Microbial Metal and Metalloid Metabolism: Advances and Applications, ASM Press, Washington, DC, pp. 297–319.

Perkins W.T. (2011) Extreme selenium and tellurium contamination in soils - An eighty year-old industrial legacy surrounding a Ni refinery in the Swansea Valley. Sci Total Environ 412–413:162–169.

Prange A., Birzele B., Hormes J., Modrow H. (2005) Investigation of different human pathogenic and food contaminating bacteria and moulds grown on selenite/selenate and tellurite/tellurate by X-ray absorption spectroscopy. Food Control 16:723–728.

Qin D., Tao H., Cao Y. (2008) Controlled Synthesis of Se/Te Alloy and Te Nanowires. Chinese J Chem Phys 20(6):670–674.

Ramadan S.E., Razak A.A., Ragab A.M., El-Meleigy M. (1989) Incorporation of tellurium into amino acids and proteins in a tellurium-tolerant fungi. Biol Trace Element Resear 20(3):225–232.

Rasband W.S. ImageJ. U.S. National Institute of Health, Bethesda, Maryland, USA. Retrieved from http://imagej.nih.gov/ij/ Accesed on 10 November 2015.

Reinoso C., Auger C., Appanna V.D., Vásquez C.C. (2012) Tellurite-exposed Escherichia coli exhibits increased intracellular α-ketoglutarate. Biochem Biophys Resear Comm 421(4):721–726.

Sadtler B., Burgos S.P., Batara N.A., Beardslee J.A., Atwater H.A., Lewis N.S. (2013) Phototropic growth control of nanoscale pattern formation in photoelectrodeposited Se-Te films. Proc Natl Acad Sci USA 110(49):19707–19712.

Saraswathy A., Hallberg R. (2005) Mycelial pellet formation by *Penicillium ochrochloron* species due to exposure to pyrene. Microbiol Resear 160(4):375–383.

Soda S., Kashiwa M., Kagami T., Kuroda M., Yamashita M., Ike M. (2011) Laboratory-scale bioreactors for soluble selenium removal from selenium refinery wastewater using anaerobic sludge. Desalination 279(1-3):433–438.

Sridharan K., Ollakkan M.S., Philip R., Park T.J. (2013) Non-hydrothermal synthesis and optical limiting properties of one-dimensional Se/C, Te/C and Se-Te/C core-shell nanostructures. Carbon 63:263–273.

Taylor D.E. (1999) Bacterial tellurite resistance. Trends Microbiol 7:111–115.

Tien M., Kirk T. (1988) Lignin peroxidase of *Phanerochaete chrysosporium*. In: Methods in enzymology. Biomass, Part B: lignin, pectin and chitin, Wood and Kellog (Eds.). Academic Press, San Diego, CA, pp. 238–249.

Tripathi K., Bahishti A.A., Majeed-Khan M.A., Husain M., Zulfequar M. (2009) Optical properties of selenium-tellurium nanostructured thin film grown by thermal evaporation. Phys B: Condensed Matter 404(16):2134–2137.

Turner R.J., Borghese R., Zannoni D. (2012) Microbial processing of tellurium as a tool in biotechnology. Biotechnol Adv 30(5):954–963.

Wittier R., Baumgartl H., Lübbers D.W., Schügerl K. (1986) Investigations of oxygen transfer into *Penicillium chrysogenum* pellets by microprobe measurements. Biotechnol Bioeng 28:1024-1036.

Yetis U., Dolek A., Dilek F.B., Ozcengiz G. (2000) The removal of Pb (II) by *Phanerochaete chrysosporium*. Wat Res 34(16):4090–4100.

Zannoni D., Borsetti F., Harrison J.J., Turner R.J. (2008) The bacterial response to the chalcogen metalloids Se and Te. Adv Microb Physiol 53:1–71.

Zhou B., Zhu J.J. (2006) A general route for the rapid synthesis of one-dimensional nanostructured single-crystal Te, Se and Se–Te alloys directly from Te or/and Se powders. Nanotechnology 17(6):1763–1769.

CHAPTER 8

Mycotechnology for the treatment of Se and Te contaminated effluents and biomineralization of Se^0 and Te^0 nanoparticles

Abstract

The potential of fungi in environmental biotechnological applications has gained popularity in recent years, particularly for wastewater treatment and green production of nanoparticles. However, the drawbacks of fungal technology, mainly those associated with operational difficulties (*e.g.* bacterial contamination, overgrowth of biomass, high maintenance of continuous operations), limit their full scale application and leave their potential largely untapped. Moreover, the biochemical mechanisms of the mycogenic production of nanoparticles are still not well understood. This chapter provides an outlook of the current state of the use of fungi in the removal of Se and Te from wastewater and biomineralization of Se and Te nanoparticles, opening up this enthralling area and indicating the current challenges and future perspectives of these technologies. The possibility and feasibility to combine the potential of fungi to remove Se and Te from wastewater while recovering valuable nanomaterials is discussed.

Key words: selenium, tellurium, biomineralization, fungi, wastewater treatment, nanoparticles.

8.1 Mycotechnology

Fungi, a group of heterogeneous eukaryotic organisms, are well known for their ability to produce large amounts of enzymes and reductive proteins. These organisms possess a rigid cell wall, composed of glycoproteins and polysaccaharides (*e.g.* glucan and chitin), which provides mechanical strength enduring changes in osmotic pressure and environmental stress (Bowman and Free, 2011). As chemo-organotrophs, fungi have simple nutritional requirements (Holan and Volesky, 1995), and are capable to grow under neutral to mild acidic conditions (~4.5-7.0). This particular set of abilities makes fungi versatile for their application in different biotechnological applications, including food production, medicine, environmental remediation and agriculture.

Although the use of fungal technology has gained popularity in the past years, its use in full scale applications is still in its infancy. The successful application of mycotechnology requires the integration of different disciplines and expertise on fungal bioprocess which may range from introducing a single fungus in biocontrol process to the genetic engineering of the fungi to induce and enhance the production of a particular enzyme or metabolite. Mycotechology is of particular interest in the wastewater treatment and biomineralization fields. Different fungal species have been used to remove organic and inorganic compounds from polluted effluents (Espinosa-Ortiz et al., 2016; Chapter 2), showing potential for real scale applications. The mycomineralization of nanoparticles has also been explored and demonstrated to be a feasible and an eco-friendly approach for the synthesis of nanomaterials (Yadav et al., 2015). However, the drawbacks of fungal technology, mainly those associated with operational difficulties (*e.g.* bacterial contamination, overgrowth of biomass, high maintenance of continuous operations), limit their full scale application and leave their potential largely unexplored.

8.2 Fungal technology in the removal of Se and Te from wastewater

Water soluble oxyanions of selenium, Se, (selenite, SeO_3^{2-}, and selenate, SeO_4^{2-}) and tellurium, Te, (tellurite, TeO_3^{2-}, and tellurate, TeO_4^{2-}) can be anthropogenically or geologically released into the environment representing a threat to living organisms (Belzile and Chen, 2015; Lenz and Lens, 2009). These metalloids are found in agricultural and industrial effluents, mostly related to mining, refinery and electronic industries (Lenz and Lens, 2009; Perkins, 2011; Biver et al., 2015). The current discharge limits of Se for the protection of aquatic life (<0.05 μg Se L^{-1}, US EPA) are debatable, raising concerns about the safety of the set limits due to the bioaccumulation of Se. So far, there is a lack of regulation regarding the discharge of Te, which will have to be address in the near future.

The use of biological treatment to remove Se from polluted effluents has emerged as an attractive cost-effective solution (Nancharaiah and Lens, 2015). Bacteria are the preferred biological agents for wastewater treatment, although the use of fungi and algae is also promising and emerging as new treatment technologies. Most biological treatments to remove Se rely on the microbial reduction of the oxyanions to their elemental form (Se^0), which is more stable and less toxic (Winkel et al., 2012). The use of fungi for the removal of Se and Te from wastewater is advantageous compared to other biological agents: fungi are easy to handle, they can grow on simple media and under acidic conditions, which makes them ideal for the treatment of acidic or mild acidic effluents (*e.g.* acid mine drainage). Se and Te transformation pathways in filamentous fungi include uptake (reduction, accumulation and assimilation) and volatilization (Gharieb et al., 1995; Brady et al., 1996). Different fungal species have been found to be Se- and Te-reducing organisms (Table 8.1). This opens up an opportunity area of investigation considering that just a small fraction of the existing fungi in nature has been explored for their potential to reduce Se and Te species.

Table 8.1 Se and Te reducing fungal organisms.

Fungi	Se-reducing organism	Te-reducing organism	Reference
Alternaria alternata	✓		Sarkar et al., 2011
Aspergillus sp.	✓		Gharieb et al., 1995; Prange et al., 2005
Coriolus versicolor	✓		Gharieb et al., 1995
Fusarium sp.	✓	✓	Ramadan et al., 1988; Gharieb et al., 1995, 1999.
Gliocladium roseum	✓		Srivastava and Mukhopadhyay, 2015
Mortierella spp.	✓		Zieve et al., 1985
Penicillium	✓	✓	Brady et al., 1996; Gharieb et al., 1999
Phanerochaete chrysosoporium	✓	✓	Espinosa-Ortiz et al., 2015; Chapter 3 and 7
Rhizopus arrhizus	✓		Gharieb et al., 1995
Scopulariopsis brevicaulis	✓	✓	Boriová et al., 2014
Thricoderma reesii	✓		Gharieb et al., 1995

Se and Te fungal reduction, and thus Se and Te removal from wastewater, is driven by a series of factors: i) fungal strain; ii) speciation and concentration of the oxyanions in solution, *e.g. P. chrysosporium* was able to reduce selenite to Se^0 but not selenate (Chapter 3), and it was more inhibited (on a mass basis) by selenite than by TeO_3^{2-} (Chapter 8); iii) medium composition, including the carbon source and nutrients (Gharieb et al., 1995; Chapter 3); iv) operational conditions, *i.e.,* temperature and pH (Chapter 3); and v) type of fungal growth, *i.e.,* selenite reduction to Se^0 has been achieved with different type of fungal growth such as disperse mycelium, pellets and as biofilms (Gharieb et al. 1995; Chapters 3 and 7).

Although the biological reduction of Se and Te oxyanions has been shown in pure culture incubations, its full scale application is still challenging due to the water quality requirements, separation and disposition of the Se- or Te- bearing biomass/solids and the recovery of valuable products (*i.e.,* Se^0 and Te^0 nanoparticles). Different bioreactor configurations have been used for the removal and recovery of Se from polluted effluents (Table 8.2). A fungal pelleted reactor can be used as an alternative treatment for the continuous removal of selenite from acidic effluents (Espinosa-Ortiz et al., 2015b; Chapter 4). The removal efficiency of the bioreactor reached around 70% at steady-state conditions, which is considerably less than the removal efficiencies (>90%) obtained by other bioreactors using bacteria (Table 8.2). However, most of the treated effluents in such bioreactors needed a pre-conditioning step to reach nearly neutral-basic conditions (pH 6.0-8.0) for optimal performance; whereas no pre-conditioning step was needed in the fungal reactor, which was able to remove Se under mild acidic conditions (pH 4.5).

The lack of full scale application of fungal bioreactors is mainly attributed to the drawbacks associated with dispersed fungal growth (*i.e.,* poor rheology of the fungal broth, attachment of biomass on reactor walls, agitators, baffles, sampling and nutrient addition lines) (Espinosa-Ortiz et al., 2016, Chapter 2). A suitable option to overcome such disadvantages is the use of fungal pellets instead of dispersed mycelium, which will improve the culture rheology facilitating the harvesting, settleability and separation of the biomass from the treated effluent. The pelletization of the fungal biomass depends mainly on the inoculation, medium composition and cultivation conditions (Espinosa-Ortiz et al., 2016, Chapter 2). The maintenance of the pellets during long term reactor operation is not an easy task, although, among different reactor configurations, the airlift reactor seems to be the best option to overcome this limitation. Bacterial contamination is a common issue in fungal reactors when working under non-sterile conditions, which might lead to the deterioration of the fungal activity in the reactor by disrupting the fungal growth and decreasing the expression of enzymes (Libra et al., 2003; Rene et al., 2010). Different strategies can be applied to suppress bacterial growth in fungal reactors under non-sterile conditions such as promoting immobilized fungal growth, reducing the medium pH, using nitrogen-limited media, microscreening and selectively disinfecting (Espinosa-Ortiz et al., 2016; Chapter 2).

The presence of Se and Te affects the fungal morphology (Chapters 3, 7 and 8), leading to the densification of the biomass and favoring more compact and smooth pellets. This would also be an advantage in fungal reactors since denser pellets would have a better settleability, facilitating the solid-liquid separation in the reactor. Moreover, Se and Te oxyanions also had an effect on the biomass yield (Chapters 3 and 8), which would help controlling excessive biomass growth (Chapter 4), a common problem in fungal reactors. The biological treatment of Se polluted effluents usually requires a settling or filtration step to separate the produced Se^0 from the treated effluent. However, due to its colloidal nature, some Se^0 can remain in the effluent, thus requiring a post-treatment step such as coagulation (Staicu et al., 2014). The entrapment of Se^0 and Te^0 within the fungal biomass is an advantage of the use of fungal pellets for Se and Te removal (Chapters 3 and 8), allowing the easy separation of Se^0 and Te^0 from the effluent of the reactor. Fungal treatment for the removal of both Se and Te oxyanions, occurring concurrently, is also promising. Different fungal species were found to be able to reduce both metalloids (Table 8.1). *P. chrysosporium* was found to be able to reduce TeO_3^{2-} and SeO_3^{2-} concurrently (Chapter 8). Although the Se and Te removal efficiencies were not that high (<20%), this shows for the first time a new approach for the parallel removal of both metalloids in mild acidic conditions. Further research is required to explore other fungal species which have shown to be able to reduce both Se and Te oxyanions to improve the Se and Te removal capacity of fungal bioreactors.

Table 8.2 Bioreactors used for the treatment of selenium polluted effluents*.

Reactor type	Inoculum	$Se_{influent}$ (mg L^{-1})	Operational conditions	Removal efficiency	References
Chemostat reactor	*Bacillus* sp.	41.8	pH 7.8-8.0, 30°C, HRT 95.2 h	99%	Fujita et al., 2002
UASB reactor	Anaerobic granular sludge	7.9	pH 7.0, 30°C, HRT 6 h	Methanogenic conditions 99% Sulfate-reducing conditions 97%	Lenz et al., 2008
Column reactor	Sulfate-reducing bacteria	2	pH 6.3-8.2, 18-22°C, HRT 48 h	>95%	Luo et al., 2008
Suspended sludge bed reactor	Anaerobic granular sludge	1.5-3.5	pH 7.5, 30°C, HRT 24 h	60%	Soda et al., 2011
UASB reactor	Anaerobic granular sludge	1.5-3.5	pH 7.5, 30°C, HRT 24 h	>95%	Soda et al., 2011
Continuous stirred-tank reactor	Activated sludge	17.2	pH 7.0-7.5, 30°C HRT 8 h	20-70%	Jain et al., 2014
Up-flow fungal pelleted reactor	*Phanerochaete chrysosporium*	10	pH 4.5, 30°C HRT 24 h	70%	Chapter 4

*Taken from Espinosa-Ortiz et al., 2016b (Chapter 4)

8.3 Fungi as Se^0 and Te^0 nanofactories

Fungi are promising systems for the "green" production of nanomaterials. These organisms have a unique set of characteristics that makes them potential biological nanofactories (Narayanan and Sakthivel, 2010). Fungi are excellent secretors of enzymes and reductive proteins, which are capable of reducing metal/metalloid ions quickly through non-hazardous processes, allowing the formation of nanoparticles with well defined dimensions and with good monodispersity (Mukherjee et al., 2001). A protein coating has been found around the nanoparticles synthesized by fungi, which makes them highly stable and improves their longevity by avoiding their aggregation even during large periods of time (Mukherjee et al., 2008).

The formation of nanoparticles of metal compounds by fungi has been attributed to an enzymatic reduction process (Zhang et al., 2011), which can occur intra- or extracellularly (Figure 8.1). However, there is a lack of information regarding the enzymes involved in such process. The mycogenyc production of metal nanoparticles has been well documented (Yadav et al., 2015), particularly for the production of nAg and nAu. Nonetheless, the mycosynthesis of Se^0 and Te^0 nanoparticles has barely been studied (Table 8.1). Synthesis of Se^0 and Te^0 nanoparticles occurs as a result of the biological reduction of the water soluble oxyanions of Se (selenite and selenate) and Te (tellurite and tellurate). Extracellular synthesis of nSe^0 was obtained with fungal filtrates of *Alternaria alternata* (Sarkar et al., 2011) and *Gliocladium roseum* (Srivastava and Mukhopadhyay, 2015). The white-rot fungus, *Phanerochaete chrysosporium*, was able to synthesize nSe^0, nTe^0 and nSe-Te (Chapters 3 and 7). Synthesis of nSe^0 occurred intracellularly and most probably also the synthesis of nTe^0 and nSe-Te was intracellularly.

A better understanding of the mechanisms of Se and Te fungal biomineralization would lead to a more controlled process for the production of the desired nanoparticles. The size, shape and crystal structure of the mycogenic nanoparticles depend on a series of factors, including fungal strain and growth conditions (Figure 8.2). *A. alternata* synthesized nSe^0 from selenate (Sarkar et al., 2011), whereas *P. chrysosporium* could only synthesize nSe^0 from selenite (Chapter 3). The formation of nSe^0 in the shape of spheres was observed in different fungal strains (Table 8.1), as also reported in the case of bacteria (Oremland et al., 2004; Kaur et al., 2009; Dwivedi et al., 2013). Incubations with TeO_3^{2-} led to the formation of nTe^0 needle-like particles by *P. chrysosporium* (Chapter 7), similar to those reported by some bacteria (Zare et al., 2012).

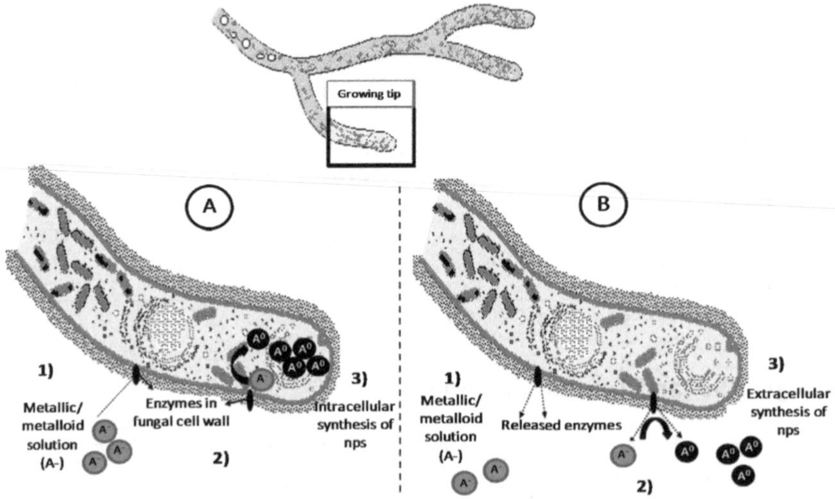

Figure 8.1 Hypothetical mechanisms of nanoparticle mycosynthesis. A -Intracellularly: 1) electrostatic interaction between metal ions and enzymes in the fungal cell wall, 2) enzymatic reduction of metal ions, 3) formation and aggregation of nanoparticles within the fungal cell. B -Extracellularly: 1) release of reductase enzyme when fungus is exposed to metal ions, 2) enzymatic reduction of metal ions, 3) formation and aggregation of nanoparticles in solution.

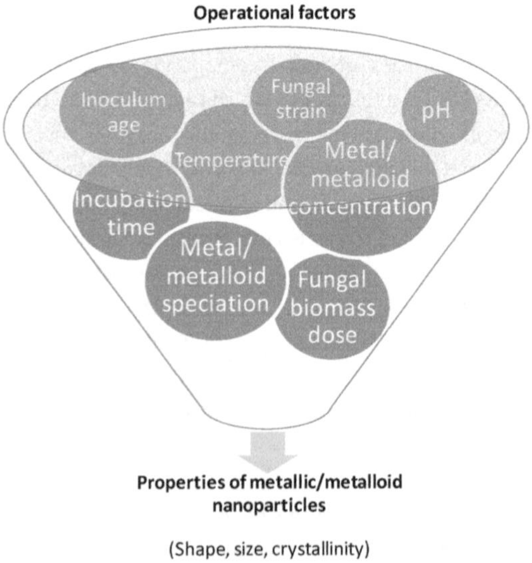

Figure 8.2 Operational factors that influence the mycogenic synthesis of metallic/metalloid nanoparticles.

P. chrysosporium was able to mineralize Se-Te nanoparticles, with different morphologies including spheres and needle like structures, when concurrently incubated in presence of SeO_3^{2-} and TeO_3^{2-} (Chapter 7). Se and Te were intimately colocalized in the Se-Te nanoparticles, however, the Se/Te ratio was different in the two morphotypes; qualitatively, the spheres had a higher content of Se, whereas the needle-like structures possessed a higher content of Te (Chapter 7). The biomineralization of Se-Te nanospheres had only been reported using bacteria (Pearce et al., 2011; Bajaj and Winter, 2014), to the best of our knowledge, this is the first time that Se-Te needle-like nanoparticles have been biologically synthesized. Physico-chemical methods have been used for the production of Se-Te nanoparticles (Qin et al., 2008; Salah et al., 2013; Fu et al., 2015); however, the biomineralization of Se-Te nanoparticles is largely untapped. The production of metal/metalloid alloy nanomaterials has attracted a lot of attention in recent years due to their enhanced properties compared to their pure nanomaterials: Se-Te nanoparticles have better electrical resistance and magnetoresistance compared to single Te and Se nanoparticles (Sadtler et al., 2013) with potential uses for the development of optical disks and other semiconducting devices (Salah et al., 2013).

Table 8.3 Mycogenic production of Se and Te nanoparticles.

Organism	Synthesis	Nanoparticle	Size (nm)	Shape	Crystal structure	References
Alternaria alternata	Extracellular	nSe^0	30-150	Spheres	Amorphous	Sarkar et al., 2011
Gliocladium roseum	Extracellular	nSe^0	20-80	Spheres	Hexagonal	Srivastava and Mukhopadhyay, 2015
Phanerochaete chrysosoporium	Intracellular	nSe^0	30-400	Spheres	Amorphous	Espinosa-Ortiz et al., 2015a (Chapter 3)
		nTe^0	20-460	Needle-like	Hexagonal	Chapter 7
		nSe-Te	50-600	Needle-like/ spheres		Chapter 7

Recovery of the produced nanoparticles is something to consider. When produced extracellularly, there should be an extra step for the collection and separation of the nanoparticles from the aqueous solution. When produced intracellularly, although the separation from the solution is very easy, a protocol to extract the nanoparticles from the biomass needs to be developed. Different fungal cell disruption methods have been proposed (Klimek-Ochab et al., 2011), including mechanical and chemical methods. For the extraction of nanoparticles trapped within the biomass a mechanical disruption method (e.g. sonication or shaking with glass beads) is proposed in order to avoid using chemicals that might affect the characteristics of the mycogenic nanoparticles.

8.4 Novel hybrid fungal sorbents containing nanoparticles for wastewater treatment

The development of new technologies and materials for the removal of pollutants has attracted the attention of researchers for several years. The creation of hybrid novel materials has emerged as an alternative for the removal of heavy metals from polluted effluents. A hybrid sorbent combines two previously known sorbents, which can complement each other to have a better efficiency, either by: i) enhancing the sorption properties of one of the sorbents by the presence of the other, or by ii) adding up their adsorption capacities. New hybrid biosorbents are developed using filamentous fungi, which are well known for their high tolerance towards heavy metals and their use as sorption materials is well documented (Viraraghavan and Srinivasan, 2011). The use of nanoparticles as sorbents is also promising, due to their enhanced properties compared to their bulk materials and high activities, including large surface areas and high reactivities (Hua et al., 2012). Development of hybrid sorbents combining fungi and nanoparticles for the removal of heavy metals has been investigated (Xu et al., 2012; Chapter 5).

The fungus *P. chrysosporium* has been used for the development of hybrid sorbents for the removal of heavy metals under mild acidic conditions. Xu et al. (2012) showed the feasibility of using iron oxide nanoparticles and Ca-alginate immobilized in *P. chrysosporium* for the removal of Pb at pH 5.0. Whereas, in Chapter 5, elemental selenium nanoparticles immobilized in pellets of *P. chrysosporium* were used for the removal of Zn at pH 4.5. The presence of nSe^0 enhanced the sorption capacity of the fungal pellets, mostly attributed to an increased on the negative surface charge density in the biomass. Compared to other biosorbents, the zinc sorption capacity of nSe^0-pellets is not that high. However, one of the main advantages of nSe^0-pellets as biosorbents is the possibility to be used under mild acidic conditions. Although the use of hybrid fungal pellets containing nanoparticles for the removal of heavy metals from acidic polluted effluents is promising, it is still in its primitive stage. Further research needs to be done regarding the reusability of the biosorbents as well as their application in continuous reactor systems.

8.5 Fungal technology for the removal of Se and potential applications of the Se^0 nanoparticles immobilized in fungal pellets

Diverse biotechnologies based on bioreduction have been used for the removal of Se oxyanions along with water reuse and Se recovery. This dissertation proposes a new technology based on the use of *P. chrysosporium* as a Se-reducing organism in a fungal pelleted reactor for the removal of selenite from a mild acidic aqueous solution, allowing the easy separation of the nSe^0 bearing pellets and their further application in other areas (Figure 8.3). The effect of selenite on the fungal morphology would enhance the settleability of the pellets, thus facilitating the solid-liquid separation in the reactor (Chapters 3 and 4). The separated fungal pellets containing nSe^0 can then: i) be reused as sorbent materials for the removal of heavy metals such as zinc (Chapter 5), or ii) be disposed for a post-treatment (mechanical disruption method, *e.g.* sonication or shaking with glass beads) to extract the synthesized nSe^0, which could in turn be used in different nanotechnology applications such as semiconductors or optoelectronics. The discovery of *P. chrysosporium* as a Te-reducing organism (Chapter 7), may suggest the use of a similar system for the removal and recovery of Te.

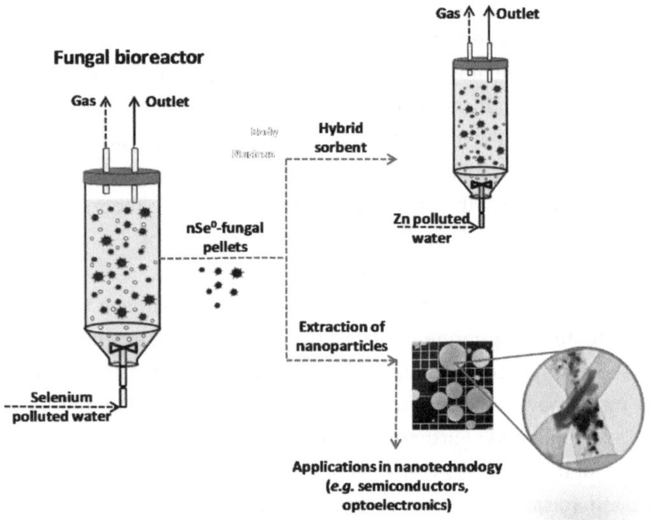

Figure 8.3 Fungal technology for removal of Se polluted wastewater and potential applications of nSe^0 immobilized in fungal pellets.

8.6 Conclusions and future perspectives

The biological reduction of Se and Te oxyanions to their elemental form (Se^0 and Te^0) allows removing the toxic Se and Te forms from wastewater and offers the possibility to recover and reuse Se and Te biominerals. The mycomineralization was shown to be a suitable technology for removing Se and Te from mild acidic solutions producing Se^0 and Te^0 particles in the nano size range. However, the application of real scale fungal bioreactors is still challenging, due to the operational difficulties associated to fungal systems (*e.g.,* bacterial contamination, fungal overgrowth, long time operation). One of the main drawbacks related to fungal reactors is the need for sterile conditions since sterilization of wastewater is not suitable. Thus, it is paramount to study the development and application of symbiotic fungal–bacterial consortiums for the removal of pollutants from wastewater. Alongside these developments, it is imperative to perform further research on post-treatment technologies, as a polishing step, to remove the residual organic matter, which is usually found at high concentrations in the treated effluent in fungal reactors. The use of fungal pellets solves the problem of separating the biomineralized species from the treated effluent, which is a common issue in bacterial reactors due to the colloidal nature of the synthesized biocomposites, but it requires a post-treatment for the extraction and recovery of the biominerals from the fungal biomass.

The use of fungal technology for the biomineralization of Se and Te nanoparticles is promising. However, there is a big gap of information regarding the mechanisms involved, such as the role of enzymes that can catalyze specific reactions leading to the synthesis of particular nanocomposites. The future of mycogenic production of nanoparticles requires strong integration with molecular techniques that allow a complete characterization of the gene expression, including genomics, proteomics and metabolomics. Furthermore, directed evolution of fungi would allow to improve their enzyme production, activity and specificity, which would allow to modify and enhance particular mycosynthetic pathways towards a more controlled production of nanomaterials. Future work will require exploring the ecotoxicology and environmental chemistry of the mycogenic produced nanoparticles. Although the routinely detection and quantification of nanoparticles at environmental concentrations is not yet performed, their release into the environment raises concerns, particularly due to the lack of information on their ecotoxicological effects. Understanding the surface chemistry of the nanoparticles, such as the nature of the capping agent, would be highly important to better understand the formation, properties and fate of such particles in the environment.

Considering that only a small fraction of the existing fungi has been tested for their ability to biomineralize Se and Te, further priorities of investigation will include extending the range of Se- and Te-reducing fungal species. This could lead to the discovery of fungal species with high tolerance to Se and Te, enhancing the removal efficiency of the wastewater treatment and probably leading to the formation of unique biominerals. Furthermore, different types of fungal growth should also be investigated for their metalloid removal efficiency and the nanomaterials produced. Depending the type of fungal growth (disperse mycellium, pellets and biofilm) certain enzymes can be better expressed or in higher quantities, which in turn can influence the biomineralization performance.

Overall, the use of fungal pellets for the treatment of Se- and Te-polluted wastewaters, as well as for the mycosynthesis of nanoparticles is an emerging interesting research area, which can have a significant direct impact on further advances in biotechnology.

8.7 References

Bajaj M., Winter J. (2014) Se (IV) triggers faster Te (IV) reduction by soil isolates of heterotrophic aerobic bacteria: formation of extracellular SeTe nanospheres. Microb Cell Fact 13:1–10.

Belzile N., Chen Y.W. (2015) Tellurium in the environment: A critical review focused on natural waters, soils and airborne particles. Appl Geochem 63:83–92.

Biver M., Quentel F., Filella M. (2015) Direct determination of tellurium and its redox speciation at the low nanogram level in natural waters by catalytic cathodic stripping voltammetry. Talanta 144:1007–1013.

Boriová K., Čerňanský S., Matúš P., Bujdoš M., Šimonovičová A. (2014) Bioaccumulation and biovolatilization of various elements using filamentous fungus *Scopulariopsis brevicaulis*. Lett Appl Microbiol 59:217–223.

Bowman S.M., Free J.F. (2011) The structure and synthesis of the fungal cell-wall. BioEssays 28:799–808.

Brady J.M., Tobin J.M., Gadd G.M. (1996) Volatilization of selenite in aqueous medium by a *Penicillium* species. Mycol Res 100:955–961.

Dwivedi S., AlKhedhairy A.A., Ahamed M., Musarrat J. (2013) Biomimetic synthesis of selenium nanospheres by bacteria JS-11 and its role as a biosensor for nanotoxicity assessment: a novel Se-bioassay. PLoS One 8:e57404.

Espinosa-Ortiz E.J., Gonzalez-Gil G., Saikaly P.E., van Hullebusch E.D., Lens P.N.L. (2015a) Effects of selenium oxyanions on the white-rot fungus *Phanerochaete chrysosporium*. Appl Microbiol Biotechnol 99:2405–2418.

Espinosa-Ortiz E.J., Rene E.R., van Hullebusch E.D., Lens P.N.L. (2015b) Removal of selenite from wastewater in a *Phanerochaete chrysosporium* pellet based fungal bioreactor. Int Biodeterior Biodegradation 102:361–369.

Espinosa-Ortiz E.J., Rene E.R., Pakshirajan K., van Hullebusch E.D., Lens P.N.L. (2016) Fungal pelleted reactors in wastewater treatment: applications and perspectives. Chem Eng J 283:553–571.

Fu S., Cai K., Wu L., Han H. (2015) One-step synthesis of high-quality homogenous Te/Se alloy nanorods with various morphologies. Cryst Eng Comm 17:3243–3250.

Gharieb M.M., Wilkinson S.C., Gadd G.M. (1995) Reduction of selenium oxyanions by unicellular, polymorphic and filamentous fungi: cellular location of reduced selenium and implication for tolerance. J Ind Microbiol 14:300–311.

Gros M., Cruz-Mortao C., Marco-Urrea E., Longrée P., Singer H., Sarrá M., Hollender J., Vicent T., Rodriguez-Mozaz S., Barceló D. (2014) Biodegradation of the X-ray contrast agent iopromide and the fluoroquinolone antibiotic ofloxacin by the white rot fungus *Trametes versicolor* in hospital wastewaters and identification of degradation products. Water Res 60:228–241.

Holan Z.R., Volesky B. (1995) Accumulation of cadmium, lead and nickel by fungal and wood biosorbents. Appl Biochem Biotechnol 53:133–146.

Jain R., Matassa S., Singh S., van Hullebusch E.D., Esposito G., Farges F., Lens P.N.L. (2014) Reduction of selenite to elemental selenium nanoparticles by activated sludge under aerobic conditions. J Environ Sci.

Kaur G., Iqbal M., Singh-Bakshi M. (2009) Biomineralization of fine selenium crystalline rods and amorpous spheres. J Phys Chem C 113:13670–13676.

Lenz M., Lens P.N.L. (2009) The essential toxin: The changing perception of selenium in environmental sciences. Sci Total Environ 407:3620–3633.

Libra J.A., Borchert M., Banit S. (2003) Competition strategies for the decolorization of a textile-reactive dye with the white-rot fungi *Trametes versicolor* under non-sterile conditions. Biotechnol Bioeng 82:736–744.

Nancharaiah Y.V., Lens P.N.L. (2015) Selenium biomineralization for biotechnological applications. Trends Biotechnol 33(6):323–330.

Oremland R.S., Herbel M.J., Blum J.S., Langley S., Beveridge T.J., Ajayan P.M., Sutto T., Ellis A., Curran S. (2004) Structural and spectral features of selenium nanospheres produced by Se-respiring bacteria. Appl Environ Microbiol 70:52–60.

Pearce C.I., Baesman S.M., Blum J.S., Fellowes J.W., Oremland R.S. (2011) Nanoparticles formed from microbial oxyanion reduction of toxic group 15 and group 16 metalloids. In: Microbial Metal and Metalloid Metabolism: Advances and Applications, Stolz J.F. and Oremland R.S. (Eds.), ASM Press, Washington, DC, pp. 297–319.

Perkins W.T. (2011) Extreme selenium and tellurium contamination in soils - An eighty year-old industrial legacy surrounding a Ni refinery in the Swansea Valley. Sci Total Environ 412–413:162–169.

Prange A., Birzele B., Hormes J., Modrow H. (2005) Investigation of different human pathogenic and food contaminating bacteria and moulds grown on selenite/selenate and tellurite/tellurate by X-ray absorption spectroscopy. Food Control 16:723–728.

Qin D., Tao H., Cao Y. (2008) Controlled Synthesis of Se/Te Alloy and Te Nanowires. Chinese J Chem Phys 20:670–674.

Rene E.R., Veiga M.C., Kennes C. (2010) Biodegradation of gas-phase styrene using the fungus Sporothrix variecibatus: Impact of pollutant load and transient operation. Chemosphere 79:221–227.

Sadtler B., Burgos S.P., Batara N.A., Beardslee J.A., Atwater H.A., Lewis N.S. (2013) Phototropic growth control of nanoscale pattern formation in photoelectrodeposited Se-Te films. Proc Natl Acad Sci 110(49):19707–19712.

Salah N., Habib S.S., Memic A., Alharbi N.D., Babkair S.S., Khan Z.H. (2013) Syntheses and characterization of thin films of $Te_{94}Se_6$ nanoparticles for semiconducting and optical devices. Thin Solid Films 531:70–75.

Sarkar J., Dey P., Saha S., Acharya K. (2011) Mycosynthesis of selenium nanoparticles. Micro Nano Lett 6:599.

Srivastava N., Mukhopadhyay M. (2015) Biosynthesis and structural characterization of selenium nanoparticles using *Gliocladium roseum*. J Clust Sci 26:1473–1482.

Staicu L., van Hullebusch E., Lens P.N.L. (2014) Electrocoagulation of colloidal biogenic selenium. Environ Sci Pollut Res 22(4):3127–3137.

Viraraghavan T., Srinivasan A. (2011) Fungal biosorption and biosorbents. In: Kotrba P, Mackova M, Macek T (eds) Microbial biosorption of metals. Springer, The Netherlands, pp. 143–158.

Xu P., Zeng G.M., Huang D.L., Lai C., Zhao M.H., Wei Z., Li N.J., Huang C., Xie G.X. (2012) Adsorption of Pb(II) by iron oxide nanoparticles immobilized *Phanerochaete chrysosporium*: Equilibrium, kinetic, thermodynamic and mechanisms analysis. Chem Eng J 203:423–431.

Yadav A., Kon K., Kratosova G., Duran N., Ingle A.P., Rai M. (2015) Fungi as an efficient mycosystem for the synthesis of metal nanoparticles: progress and key aspects of research. Biotechnol Lett 37:2099–20120.

Zare B., Faramarzi M.A., Sepehrizadeh Z., Shakibaie M., Rezaie S., Shahverdi A.R. (2012) Biosynthesis and recovery of rod-shaped tellurium nanoparticles and their bactericidal activities. Mater Res Bull 47:3719–3725.

Zhang X., Yan S., Tyagi R.D., Surampalli R.Y. (2011) Synthesis of nanoparticles by microorganisms and their application in enhancing microbiological reaction rates. Chemosphere 82:489–494.

Zieve R., Ansell P.J., Young T.W.K., Peterson P.J. (1985) Selenium volatilization by *Mortierella* species. Trans Br Mycol Soc. 84:177–179.

Appendix 1

Nomenclature

C_i	Initial concentration of adsorbate in solution (mg L^{-1})
C_e	Concentration of adsorbate in solution at equilibrium (mg L^{-1})
q_e	Amount of adsorbate adsorbed at equilibrium (mg g^{-1})
q_m	Maximum sorption capacity (mg g^{-1})
$q_{e\ cal}$	Calculated sorption equilibrium capacity (mg g^{-1})
$q_{e\ exp}$	Experimental sorption equilibrium capacity (mg g^{-1})
k_L	Langmuir isotherm constant (L mg^{-1})
k_f	Freundlich isotherm constant (mg g^{-1})(L $mg^{-1})^{1/n}$
n	Exponent in Freundlich isotherm
k_1	Pseudo-first order sorption rate constant (min^{-1})
k_2	Pseudo-second order sorption rate constant (g mg^{-1} min^{-1})
h	Initial sorption rate in pseudo-second model (mg g^{-1} min^{-1})
nSe^0-pellets	Elemental selenium nanoparticles immobilized *P. chrysosporium* pellets
R^2	Correlation coefficient

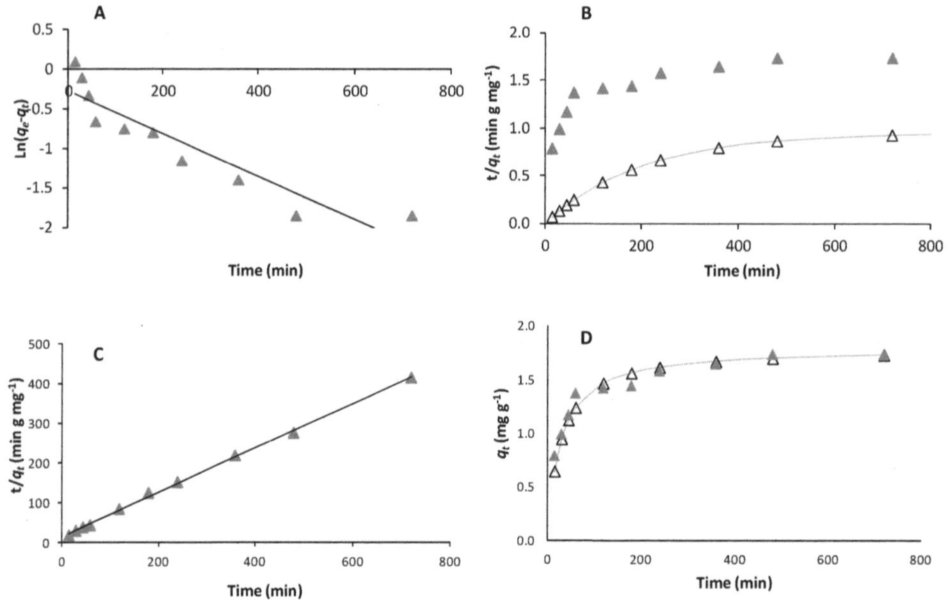

Figure A1. Sorption kinetics for Se-free pellets. A) Linearized pseudo-first order. B) Pseudo-first order model C) Linearized pseudo-second order. D) Pseudo-second order model. Experimental data ( ), Calculated data ( ).

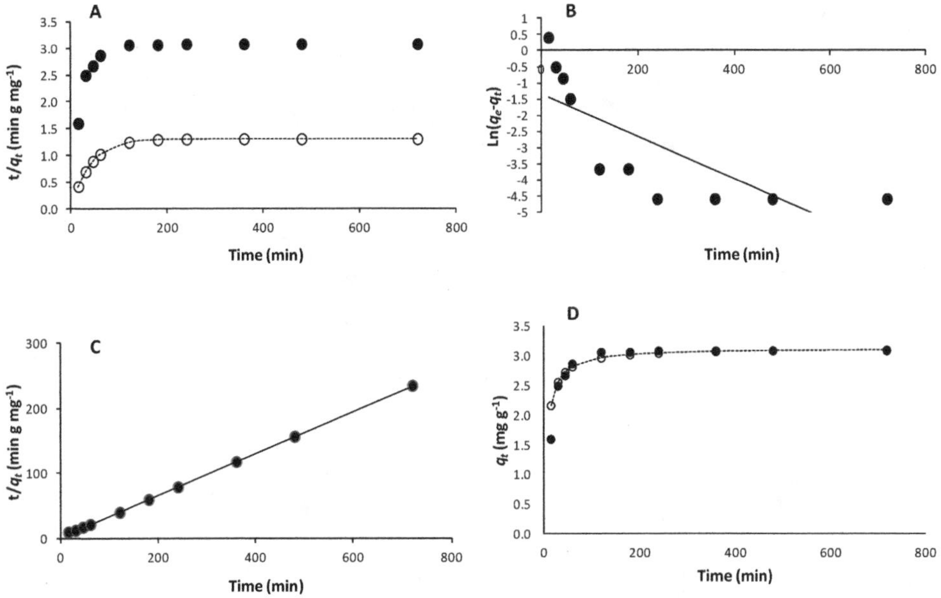

Figure A2. Sorption kinetics for Se-free pellets. A) Linearized pseudo-first order. B) Pseudo-first order model C) Linearized pseudo-second order. D) Pseudo-second order model. Experimental data ( ), calculated data ( ).

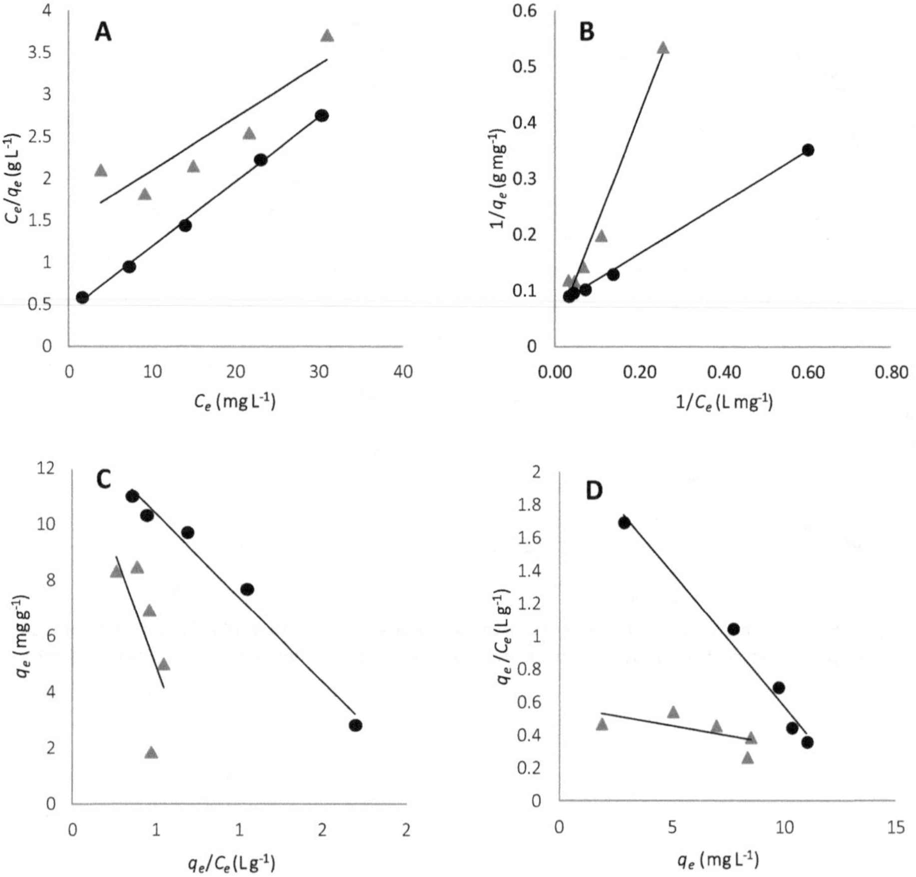

Figure A3. Linearized Langmuir isotherms for Zn sorption. A) Type I Hanes-Woolf, B) Type II Burke, C) Type III Eadie-Hofstee, and D) Type IV Scatchard. Se-free (▲), nSe^0-pellets (●).

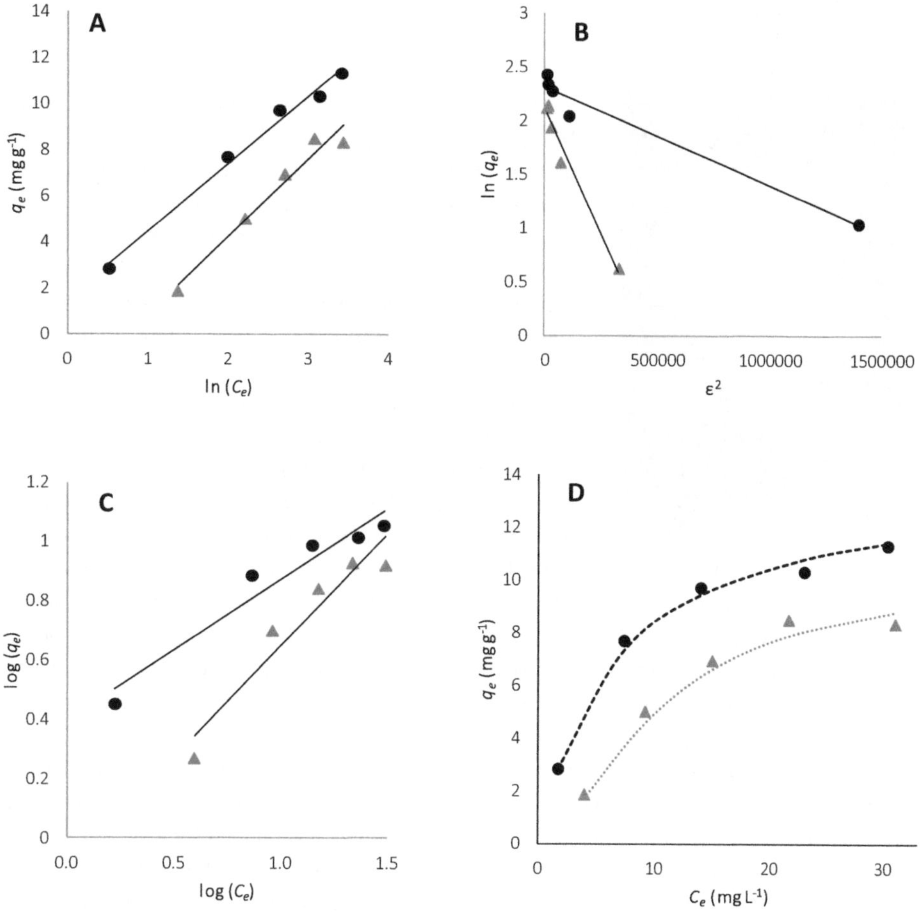

Figure A4. Linearized Temkin (A), Dubini-Radushkevich (B) and Freundlich (C) and non-linearized Sips (Langmuir-Freundlich) (D) isotherms for Zn adsorption. Se-free (▲), nSe^0-pellets (●).

Biography

Erika Jimena Espinosa-Ortiz was born on July 8th, 1984, in Mexico City. At the age of 17, Erika was selected as one of the participants to represent Mexico in the prestigious contest *Stockholm Junior Water Prize*, during the World Water Week in Stockholm, Sweeden, 2001. Being part of this experience impulsed her to pursue a career in Environmental Science. She received her Bachelor's Science degree, *summa cum laude*, in Environmental Engineering from the Autonomous Metropolitan University (UAM), Mexico, in 2006. Upon graduation, she worked for one year as environmental consultant, analyst and laboratory assistant in the Water Quality and Residues Laboratory at UAM. Erika obtained two Master's degrees. The first one in Environmental Engineering, from the National Autonomous University of Mexico (UNAM) in 2009; during this period she investigated the biological contamination of source water supplies in the Tula Valley, Mexico, which is a region that has been irrigated with wastewater for over a hundred years. She was awarded with a Fulbright scholarship to complete the second Master's, which she obtained from Purdue University, USA, in 2010 in Ecological Sciences and Engineering. Her research back then consisted on assessing the ecotoxicological effects of gallium and indium on soil microbial activities and plants. In 2012, Erika started her PhD program at UNESCO-IHE, as part of an Erasmus Mundus Joint Doctorate Program on Environmental Technologies for Contaminated Solids, Soils and Sediments (ETeCoS^3). Erika devoted her work to investigate the use of fungi as selenium and tellurium reducing organisms and their potential in wastewater treatment and nanotechnology. As part of her program, Erika also performed research at Paris-Est University and at the Center for Biofilm Engineering in Bozeman, Montana, USA. Erika has 9 years of laboratory experience and research-related work. She also has participated in international conferences and already has a number of scientific publications including a book chapter and peer-reviewed journals. Erika continues to be an active young professional with an interest to pursue a career in science.

Publications

E.J. Espinosa-Ortiz, E.R. Rene, F. Guyot, E.D. van Hullebusch, P.N.L. Lens. (2016). Biomineralization of tellurium and selenium-tellurium nanoparticles by the white-rot fungus *Phanerochaete chrysosporium*. Submitted.

E.J. Espinosa-Ortiz, M. Shakya, R. Jain, E.R. Rene, E.D. van Hullebusch, P.N.L. Lens. (2016). Use of elemental selenium nanoparticles immobilized fungal pellets fungal as sorbent material to remove zinc from water. Submitted.

E.J. Espinosa-Ortiz, Y. Pechaud, E. Lauchnor, E.R. Rene, R. Gerlach, B.M. Peyton , E. D. van Hullebusch, P.N.L. Lens. (2016). Effect of selenite on the morphology and respiratory activity of *Phanerochaete chrysosporium* biofilms. Bioresource Technology. 210:138-145.

E.J. Espinosa-Ortiz, E.R. Rene, K. Pakshirajan, E.D. van Hullebusch, P.N.L. Lens. (2016). Fungal pelleted reactors in wastewater treatment: applications and perspectives. Chemical Engineering Journal. 283:553-571.

E.J. Espinosa-Ortiz, E.R. Rene, E.D. van Hullebusch, P.N.L. Lens. (2015). Removal of selenite from wastewater in a *Phanerochaete chrysosporium* pellet based fungal bioreactor. International Biodeterioration and Biodegradation. 102:361-380.

E.J. Espinosa-Ortiz, G. Gonzalez-Gil, P.E. Saikaly, E.D. van Hullebusch, P.N.L. Lens. (2015). Effects of selenium oxyanions on the white-rot fungus *Phanerochaete chrysosporium*. Applied Microbiology and Biotechnology. 99(5):2405-2418.

E.J. Espinosa-Ortiz, M. Vaca-Mier. (2013). Nanotechnology for water and wastewater treatment: potentials and limitations. In: P.N.L. Lens, J. Virkutyte, V. Jegatheesan, S.H. Kim, S. Al-Abed (Eds.), Nanotechnology for Water and Wastewater Treatment. IWA Publishing, London, UK, pp. 83-127.

Conferences

E.J. Espinosa-Ortiz, M. Shakya, E.R. Rene, E.D. van Hullebusch, P.N.L. Lens. (2015). *Biosorption of Zn with elemental selenium nanoparticles immobilized fungal pellets of Phanerochaete chrysosporium.* Proceedings of the 4th International Conference on Research Frontiers in Chalcogen Cycle Science & Technology, Delft, The Netherlands, May 28-29, pp. 161-170.

E.J. Espinosa-Ortiz, E.R. Rene, E.D. van Hullebusch, P.N.L. Lens. (2014) *Exploiting the operational advantages of Phanerochaete chrysosporium inoculated suspended growth bioreactor for the removal of selenite from wastewater.* 2nd International Conference on Recycling and Reuse, Istanbul, Turkey, June 4-6.

E.J. Espinosa-Ortiz, P.N.L. Lens. (2013) Selenium stress in fungi: potential applications. Proceedings of the 3rd International Conference on Research Frontiers in Chalcogen Cycle Science & Technology, Delft, The Netherlands, May 27-28, pp. 31-38.

SENSE

Netherlands Research School for the
Socio-Economic and Natural Sciences of the Environment

D I P L O M A

For specialised PhD training

The Netherlands Research School for the
Socio-Economic and Natural Sciences of the Environment
(SENSE) declares that

Erika Jimena Espinosa-Ortiz

born on 8 July 1984 in Mexico City, Mexico

has successfully fulfilled all requirements of the
Educational Programme of SENSE.

Paris, 10 December 2015

the Chairman of the SENSE board

Prof. dr. Huub Rijnaarts

the SENSE Director of Education

Dr. Ad van Dommelen

The SENSE Research School has been accredited by the Royal Netherlands Academy of Arts and Sciences (KNAW)

The SENSE Research School declares that Ms Erika Espinosa-Ortiz has successfully fulfilled all requirements of the Educational PhD Programme of SENSE with a work load of 44.9 EC, including the following activities:

SENSE PhD Courses

- Environmental Research in Context (2013)
- Speciation and bioavailability (2013)
- Research in Context Activity: 'Co-organising ETeCoS3 Summer School (Characterization and Remediation of Contaminated Sediments) and compiling Abstract Book, Delft, The Netherlands' (2013)

Other PhD and Advanced MSc Courses

- ETeCoS3 PhD Introductory Course, University of Cassino, Italy (2014)
- Contaminated soils and remediation, University of Paris-Est, France (2015)
- Communicating water - Bridging the Gap between Science and Society, UNESCO-IHE Delft, The Netherlands (2015)

External training at a foreign research institute

- Training on 'The use of oxygen microelectrodes', UNISENSE Company, Denmark (2015)
- Training on 'Confocal and optical microscope for transmitted light and epi-fluorescent imaging, and the use of cryosection equipment', Center for Biofilm Engineering, United States (2015)

Management and Didactic Skills Training

- Supervising Intern with project entitled 'Influence of operational parameters on the effects of selenite on fungal pellets of *Phanerochaete chrysosporium*' (2013)
- Supervising MSc students with thesis entitled 'Biosorption of Zn with elemental selenium nanoparticles immobilized fungal pellets of *Phanerochaete chrysosporium*' (2014)

Oral Presentations

- *Mycogenic production of elemental selenium nanoparticles*. UNESCO-IHE PhD Symposium, 23-24 September 2013, Delft, The Netherlands
- *Exploiting the operational advantages of Phanerochaete chrysosporium inoculated suspended growth bioreacttor for the removal of selenite from wastewater*. 2nd International Conference on Recycling and Reuse, 2-5 June 2014, Istanbul, Turkey

SENSE Coordinator PhD Education

Dr. ing. Monique Gulickx

* 9 7 8 1 1 3 8 0 3 0 0 4 6 *

T - #0416 - 101024 - C44 - 244/170/10 - PB - 9781138030046 - Gloss Lamination